MW00484545

What's Left of the World

Education, Identity and the Post-Work
Political Imagination

What's Left of the World

Education, Identity and the Post-Work Political Imagination

David J. Blacker

Winchester, UK
Washington, USA

First published by Zero Books, 2019
Zero Books is an imprint of John Hunt Publishing Ltd., No. 3 East St., Alresford,
Hampshire SO24 9EE, UK
office1@jhpbooks.net
www.johnhuntpublishing.com
www.zero-books.net

For distributor details and how to order please visit the 'Ordering' section on our website.

Text copyright: David J. Blacker 2018

ISBN: 978 1 78904 010 4
978 1 78904 011 1 (ebook)
Library of Congress Control Number: 2018942283

All rights reserved. Except for brief quotations in critical articles or reviews, no part of this
book may be reproduced in any manner without prior written permission from the publishers.

The rights of David J. Blacker as author have been asserted in accordance with the Copyright,
Designs and Patents Act 1988.

A CIP catalogue record for this book is available from the British Library.

Design: Stuart Davies

UK: Printed and bound by CPI Group (UK) Ltd, Croydon, CR0 4YY
US: Printed and bound by Thomson-Shore, 7300 West Joy Road, Dexter, MI 48130

Also by the Author

The Falling Rate of Learning and the Neoliberal Endgame
Zero Books, 2013

We operate a distinctive and ethical publishing philosophy in
all areas of our business, from our global network of authors to
production and worldwide distribution.

Contents

For Carolyn

How do we create and proliferate a compelling vision of economies and ecologies that center humans and the natural world over the accumulation of material?
We embody. We learn. We release the idea of failure, because it's all data.
But first we imagine.
We are in an imagination battle.
adrienne marie brown (2017)[i]

Preface: Useless education

Our struggle must be towards the construction of a new and surprising world, not the preservation of identities shaped and distorted by capital.
Mark Fisher (2013)[1]

The humanist dream of a society in which education is valued intrinsically finally lies on the threshold of being achieved. Yet there will be no celebrations. There has not been a great awakening to the life of the mind or a widespread regenerative commitment to an aesthetics of existence. Or any such thing. What has happened instead is more like a rug being pulled out from under everyone's feet: the reigning narrowly instrumental attitude toward education—how does it *pay?*—is on the verge of being toppled not by a hopeful humanism but by a deflating cynicism. It is a system that was built on certain promises that came due and then defaulted. In the wake of this bubble of expectations bursting, education's intrinsic value—if it can be relocated—may be all that is left.

Due to large-scale economic and technological changes, educational institutions are no longer as useful vocationally as advertised, no longer as reliable a road to prosperity for individuals, nor so obvious a route to societal wealth. One of the great promises of modernity, epitomized by optimistic progressives like the American pragmatist John Dewey, that the development and exercise of collective intelligence will lead to cumulative moral and material progress, appears to be broken.[2] The problem was not simple failure. It was, analogous to what political theorist Amartya Sen once demonstrated about contemporary famines, primarily a problem of distribution of the promised peace and productivity dividends of the postwar periods (WWII and Cold War), including the profligate release of eons of fossilized fuels.[3] In retrospect it was cringe-worthily naïve for anyone to expect that the monstrous influx of postwar wealth would suddenly be distributed evenly among the masses out of the goodness of our overlords' hearts. This is akin to the mistake made by low information types who favor "drill baby drill!" or Middle Eastern adventurism in order to secure "our oil," as if we the people would all soon be receiving stipends from Chevron and Exxon Mobile like some kind of Alaskan oil dividend for every citizen. As is becoming clearer, the primary recipients of the great postwar surpluses seem to have been a new self-perpetuating, tech and finance-enabled aristocracy, swelling from the unprecedented wealth that was once bruited as a shared inheritance. Yet somehow, amidst the unprecedented prosperity, wealth inequality has widened to being worse than under Imperial Rome.[4] And there is no real plan for mitigation. Even economist Thomas Piketty, the great contemporary cataloger of intensifying global inequality, offers no real solution other than a chimerical "global wealth tax" that even he regards as quixotic.[5]

Aiding and abetting this gigantic productivity heist, our educational system has functioned as a giant upward-suctioning vacuum of all the "excess" wealth, funneling more and more of it up to those already at the top. As middle-men to this fiscal hoovering, our educational institutions, especially the higher education sector, mostly pay *themselves* through a gatekeeping positionality; they function as rent collectors of a receding aspirational social mobility — promised to run through an allegedly meritocratic educational sector — that now diminishes to a point of nonexistence, as a return on investment, for any but the most elite of the elite.[6] They are like greedy ferrymen on the eve of an invading army's river crossing, collecting fleeing refugees' last fares before the impending enemy advance. It is not that educational credentials — and in some cases actual skills — cannot help one get ahead. They still can. One can still

2

make it across the river ahead of the others if one has the cash. But like a vast prisoner's dilemma game-theoretical experiment, education's status as a zero-sum positional good renders very slim, at an overall societal level, the prospects for any significant amelioration. As is widely noted, in abject contrast to its nineteenth-century ideals, higher education seems now to be more clearly an engine of social *inequality* and as such diametrically opposed to the original modernist promises to the contrary. With the proper credentials one is better positioned to elbow the others out of one's way.

One thus "invests" in oneself: all hail and make way for the future *me*.

Across the board, a cruel "winner loses" logic of workplace automation has taken hold: akin to how the mechanized player piano in Kurt Vonnegut's eponymously titled novel displaces the human pianist, the productivity enhancements brought about through education have accreted into machines and systems ("fixed capital" in Marxian lingo) that reliably create wealth for the few yet displace and toss aside the many.[7] One's laptop represents a phantom army of past support staff, one's phone a former mass of human ex-switchboard operators, and so on. The utilitarian promise of modern capitalist innovation was that these displacements, while often difficult for individuals during the time they occur, would be more than made up for in the long run by the overall betterment of society that they produced in aggregate. We once heard repeatedly from a confident conventional wisdom with their libertarian cartoons about how the horse and buggy manufacturers all got new and better jobs at the emergent auto plants that had just displaced them. ("Silly old buggy makers!" we were all to laugh on cue.) This triumphalist progress narrative reigned supreme for generations in the twentieth century until, finally, toward the end of the century, in their latest and greatest technological turn, the productivity enhancements became so great and so extensive that there became less and less for the actual human beings involved to do with themselves. Those who were supposed to enjoy all these benefits (to be fair they *were* some benefits like better medical care, drinking water and rising literacy rates— no small achievements) were starting to be edged out of the loop of production entirely. More and more started to become economically purposeless.

A striking emblem of these trends is that the old Big Three of the US economy, GM, Ford and Chrysler ("as General Motors goes, so goes the nation"), at their height in the 1950s once employed nearly a million manufacturing workers in their region around Detroit, whereas today's Big Three, Apple, Google and Facebook, together employ just over 50,000 in total in Silicon Valley.[8] The hyper-productive digital economy—Facebook employees are calculated to produce $600K in profits *each*—in the long run expands its vast profits while shrinking its reliance on human labor.[9] A cruel twist of economic fate has thus caused education's very usefulness to render it obsolete for the majority of the population. Only *so many* creative, entrepreneurial innovative types are needed—and are even tolerable—by any economic system no matter how sophisticated. The growing exclusion of so many now rendered economically precarious and disposable is a de facto recognition of a lack of high-end demand.

The extra-economic knock-on effects are equally worrisome. As the imaginative few take over more of the work of the putatively unimaginative many, the latter are rendered not only unemployable but also increasingly disconnected from their societies. Geographer and social theorist David Harvey identifies in this phenomenon a rising "global alienation" ("global" in the two senses of all-encompassing of areas of life, e.g., work, neighborhood, family, politics, and also geographically).[10] In parallel, automation and other technological developments have indeed resulted in staggering rises in productivity, but these gains have been subject to an extreme upwardly-skewed maldistribution; the "automation dividend" is very real but it has been sequestered by a tiny minority of the population, those represented in protest liturgy as "the 1%." (According to the classical Marxist definition as those deriving their wealth from direct ownership of the means of production—rather than just "rich people"—the US capitalist class is actually about 2%.)[11] The remainder of humanity, that segment of the population asymptotically approaching 100%, is increasingly consigned to the category of an economically unproductive mass. These people have been rendered economically superfluous and it is important to recognize that they are no longer some relatively small isolable segment of the population, as in the extreme yet emblematic case of Nazi Germany's "useless eaters," i.e., the disabled, the first among those considered to be "life unworthy of life."[12] The way things are headed, the useless people are in *c. free economy* good time *all of us*. Which occupations can confidently say they are "safe" in the coming decades? The most optimistic among them would only be warranted a "maybe." *Everyone* is but one technological development away from obsolescence. So this very bad joke is at the expense of *all* of us: like educational lemmings marching off a cliff, we've added so much productive value to ourselves that we've thrown ourselves right out of work. (Lemmings don't actually do this but it's still a great image.)[13] In a weird irony, the educational value we've scrupulously added to ourselves now comes at the expense of the aggregate educational value of our fellow human beings in general. It was all supposed to be a grand societal "win-win" but it turns out that it's not so simple.

We do not need an apocalyptic *Terminator 3* "rise of the machines" in order to glimpse how more and more of us are replaceable and disposable—at least in the sense of our lacking a necessary economic place, where a planned obsolescence awaits not just our consumer goods but we ourselves as well. This is why the notorious Judeophobic chant of the Charlottesville quasi-Nazis, "you/Jews will not replace us!," is so chilling: the conviction underlying it is *half* true, in the sense that the wholesale replaceability of human beings is a core moral issue of our times.[14] Although its dumb racist posturing obviously causes it to err in its identification of the source of the problem (viz., "the Jews"), the alt-right is in an odd way more advanced than the identity politics left on this matter, as the left tends chronically to underestimate the extent to which even the "white males" who are, *ex hypothesi*, supposed to inhabit an easy and stable privilege-enjoying position, are in fact also existentially threatened along with everyone else. Yesteryear's concern with mass economic exploitation now becomes outdated, not because it no longer exists—it does, copiously, especially in less economically developed countries [LEDCs]—but because it is diminishing as an economic possibility in the more economically developed countries [MEDCs]. As it will eventually in the LEDCs too. What's new is that those tech entrepreneur Andrew Yang calls "normal people"—the non-hyper-credentialed—*en masse* are beginning to find themselves in the classic *lumpen* position of having *less* to offer than their labor itself. This situation represents an undercutting of the place that "the masses" should inhabit in classical Marxism, at least if they are to constitute any kind of revolutionary force.[15] They are no longer so much cogs in the machine as they are rendered external and incidental to it. As economist Joan Robinson once put it, the "misery of being exploited by capitalists is nothing compared to the misery of not being exploited at all."[16]

From the point of view of education, this situation is both worse and better than is typically imagined. It is worse in that mainstream economists' nostrums are largely beside the point. The conventional view embraced by nearly everyone except, lately, some on the far fringes who have reached an attitudinal tipping point, is that we can educate ourselves into prosperity and security.[17] (This is perhaps the one thing to appreciate about the alt-right: amidst their reflexive hostility to campus politics, they have shown a willingness to dispense with the longstanding liberal faith in the salvific powers of an educated elite.) We must race to get better, faster, smarter, etc. so that we can win the high-tech, high-wage jobs of tomorrow. Most have seen through this long con already, but its animating presumptions continue to shape the legitimating language of formal education at all levels. Especially in higher education, one must always anxiously strive to keep up with the developmental pace and to anticipate, in the American system as an 18-year-old, the jobs of the future; it is a bad break if you lose at the game of life because of your teenage wagering on the wrong college major. Judging from universities' public relations, the incredibly exciting and increasingly globalized life we are leading will only continue upwards and onwards—so long, of course, as said universities continue to be able to churn out their magically inventive graduates. Unknowingly hearkening back to the Platonic conception of knowledge as a kind of courage,[18] the presumption is that all it takes is a bootstrapping existential will heroically to grasp the future and, in the breathless but typically nihilistic slogan of one US research university, "Dare to be First!"[19]

If everyone would stop being so fearful and plunge themselves headlong into whatever new skills course is on offer, supposedly there will be more than enough jobs for everyone. We can all design phone apps or repair drones—or maybe repair the machines that repair the apps that repair the drones—and then repair *to* whatever increasingly degrading personal service crevices remain. All this might be fine were the worldwide population somehow rapidly dwindling. But it isn't. Quite the opposite. We are faced with an economic environment in which the labor intensiveness of the overall economy diminishes while the population rises. Add to this labor demand/population inversion the abovementioned upward maldistribution of wealth, and you get a recipe for a massively idle and alienated population that is simultaneously also less reliably linked into their environing communities. The central problem with any imagined educational solution to mitigate the effects of these economic pincers is that augmenting the productivity and profit-generating capacities of individuals may well, as a function of the timing of their life course, succeed for *particular* individuals (i.e., they may become among the lucky few societal "winners"), but for society *as a whole* it cannot similarly succeed because "productivity" means, in its essence, doing more with less. Though this wasn't supposed to be the case in the grand educational win-win, the "less" that can be done more with turns out to be less of *us*. It is difficult for individuals to see this big picture from within their own horizons because it still makes sense for them to pursue career paths through formal education; it is still the only sensible option for most people. It would be irresponsible, absent any unusual personal circumstances, to advise a teenager to drop out of school and to proceed uncredentialed into the labor market. This is not a situation in which there can be a societal fix based on the sagacity of individuals' educational choices. One cannot just "opt out" of the economy like one might choose to play or not play a particular game. The opportunity cost, except perhaps for a tiny number of extreme ascetic eccentrics, is simply too high. One would largely have to abandon society itself. Hopefully from within a tolerable niche, one has to play this game like it or not. Pretending otherwise is delusional.

The productive "winners" have won precisely because they have shown themselves positionally able to displace others. Now the economists will tell us with more textbooky parables about Billy's firm that made widgets more efficiently and hired brilliant managers and so got better sales and then grew and hired more workers, etc., and that's wonderful for Billy, Inc. (The allegedly ignorant who don't understand economics and these little parables think it's a zero-sum game when in fact it's one in which *everyone* prospers: Free Enterprise 101.) Yet as both higher tech automation-generating fixed capital and skills-stuffed workers accumulate and become a dime a dozen, by definition there is less and less in the traditional economic sense for the mass of human former workers to *do* as part of the value creation process. This situation may be saved a bit by better and more just distributional schemes such as Universal Basic Income strategies and/or massive social welfare and safety net spending (funded by a Pikettian global tax perhaps), but none of that will "bring back jobs" (as both the rightist Trump and leftist Sanders campaigns fatuously promised during the 2016 US presidential elections) and regenerate the retrospectively white male golden era of the fully employed American postwar generations. Such policies might bring a lot of things—nobody knows for sure—but time itself would have to reverse for them to bring back *those* blue-collar jobs. But even if, *per impossibile*, wealth were massively redistributed and inequality ameliorated this would be *on the basis* of the new economy not in replacement of it. Thus the educational question would remain just as acute: what exactly will the burgeoning mass of the economically idle *do* with themselves?

One thing seems certain, though: in such an environment the value-addedness rationale for educational expenditures, the Jeffersonian civic universalism that at least in theory has justified the expansion of US public education for nearly 2 centuries, is no longer nearly so compelling.[20] The "products" of that system are simply no longer needed on as large a scale. Hence the central educational problem becomes *what to do with all of these extra people—most of us—who have been rendered economically useless?* This book attempts to start answering that question. It turns out it has everything to do with education but little to do with formal schooling and "educational reform" or the tempests within our teapot schools or universities—these institutions

9

may be relevant or they may not. Something much deeper and more encompassing is needed. This problem is far beyond any technocratic policy fix, juridical decree, or clever school reform.

Believe it or not, the very urgency of the "useless people" question represents some potential good news: though they have grown fat and lazy on widespread belief in their economic utility and alleged egalitarianism, our schools and colleges may now be freer to remake themselves in a different image. And the really good news for those of an intellectual bent is that figuring out just what that remaking might look like is an inevitably philosophical project. *What are we going to do with ourselves?*

Yes we can dissolve ourselves into an internet hive mind and just sort of vanish as individuals into a tide of entertainment memes and shopping preferences. That will probably work for some of the population. But others will need to locate some larger meaning to their lives and overall point to their existence. It is impossible to develop a rationale for educational endeavor without some eudaimonistic baseline, a vision of the good life for a human being, to put it in Aristotelian terms. Social reproduction in the anthropological sense will continue, of course, as long as there are human beings, as children's guardians will have to make choices about what kinds of formative experiences their charges are to have. However, this relatively new — in the 200,000-year span of the epochal life and times of anatomically modern humans — set of customs and practices of *formal* education could take on a multitude of forms and be repurposed in an almost infinite variety of ways. The myopic presentism that gives undue weight to the institutional arrangements of recent centuries must yield to a longer view.

With the yoke of economic utility shucked off, should we decide to continue them, our school-type endeavors might for example be much more strongly civic in orientation, for example, a trend we may already be seeing in some areas where explicit norms of socially responsible behavior become

10

prominent (e.g., antiracism, environmentalism). Or they could be humanistic in the classical sense of promoting a specific vision of the human good, a *paideia* (see Chapter 2). Or perhaps even aesthetic or discovery oriented. They could be just about anything. But because of the large outlay of resources they require, under current conditions due largely to the legal legacy of universal education (e.g., state laws guaranteeing education to all, gender, racial and disabilities inclusion along with other civil rights, etc.), this question of how we orient these massive institutions will become more pressing and interesting than it has in over a century. With the scoundrel's justificatory refuge of economic utility removed, the point of it all actually becomes an unavoidable question, not just for a few academics but now for normal people as well. It's *not* just the economy, stupid!

In sum, while the bad news is that education's economic utility justification is no longer compelling, the good news is *also* that education's economic utility justification is no longer compelling. Like a crutch finally being removed after a long course of physical therapy, this situation is both frightening and exhilarating. One could finally walk under one's own power or one could fall flat on one's face. It is frightening because it constitutes a recognition that what we thought was solid ground beneath our feet is no longer to be trusted at all. Stay in school, study hard and get ahead. "Maybe you will get ahead, maybe you won't" is now the only honest response to this zombie mantra. A *few* will, to be sure, a *few* are still needed and will be for some time. But fewer and fewer. In the American experience this tectonic change in education's relation to the economy is all the more worrisome because it threatens to sever the long-cherished presumed linkage between educational effort and social position. This has long been a core legitimating ideal of our quasi-meritocracy: positions in society are *deserved* because of an individual's virtue, talent and hard work, etc. rather than merely guaranteed to them in the feudal sense via heredity.

Not that everyone always believed in this. In fact, those at a social hierarchy's very bottom and very top, and especially the former, are typically more perspicacious about that hierarchy's legitimating myths than are those in the mainstream for whom it is often more difficult to see the "edges." But the idea that educational attainment justifies social position is extensively ingrained in American culture, particularly, unsurprisingly, among educated elites.

However jaded critical types have become of this nexus, it is important to appreciate how unsettling it is for a great many people—particularly among the middle and upper-middle classes—to think that their privileges are not in some deep sense due to their own personal qualities, those qualities mostly consisting in what was educated into them (although in my own personal experience I have been surprised at the extent to which elites, behind closed doors and perhaps after a few drinks, admit to harboring harder-core genetically-based accounts of their supposed superiority). To think that living in a certain neighborhood and placing one's kids in "the right schools" and so on no longer guarantees anything is disturbing to say the least. Many are currently realizing that despite their advantages there is little chance that their offspring will enjoy the same level of wealth and privilege as they have. Some will be fine of course. But a great many will not. At best they can anticipate living off the remaining fumes of accumulated family wealth (including real estate, paid off mortgages and the like), although this tends to run out fairly quickly for most. What the successor generation cannot anticipate is *making their own way* on anything like the terms of their parents and grandparents' generations. It looks like a big lie to them, this idea that they'll secure themselves through their educational performances. As for the aspiring lower orders, to the extent that they believed the meritocratic edu-myths in the first place, what made the obvious structural unfairness at least tolerable, the whole sense that despite all there persisted an orderly pathway out for the willing and able, seems to come undone. This is a situation that German philosopher Jurgen Habermas famously identified as a "legitimation crisis," where institutions are unable to achieve their official goals.[21]

The exhilarating part for fans of political change is that legitimation crises require resolutions. An optimistic scenario is that the aims and purposes of education will once again take center stage and become reanimated as real and vibrant and even urgent public questions. Historically, education has been very good at forcing such questions about itself, a preferred venue for them in fact. Though social reproduction is ongoing and slow motion and long-term, education always has a "fierce urgency of now" for current parents who cannot wait a decade because by then their charges' childhood is gone.[22] This urgency can force philosophical questions, as it has during previous periods such as the emergency of the common school during the early and mid-nineteenth century. It is not that everyone suddenly becomes philosophical and versed in Aristotelian means-ends talk, but as the collective need emerges to re-chart a course for these institutional behemoths, basic assumptions may get re-examined and long-dormant discussions may reanimate. The ideologically soporific power of education's economic utility justification has been so pervasive that the needed wakeup process promises to be very powerful. Without economic utility pre-empting every conversation and restricting every horizon there could be a mini-philosophical renaissance where alternative aims and purposes are considered. Perhaps, against all odds, an aesthetic basis for education might once again be taken out of the exclusive province of elites and be more widely considered, qua an emphasis on human innovation, creativity and the appreciation of the same. (A long shot, granted.) Or maybe civic aims could once again become central? In the wake of Trump's "low information voters" ("I love the poorly educated!") and the repetitive invocations of "fake news," this seems indeed an urgent question.[23]

It is a question, in fact, that throws the entire set of Jeffersonian premises of democratic education into question. Have we the people succumbed to an "idiocracy" scenario and simply become too stupid and short-attentioned to be properly sovereign?[24] Have we lost that particular world historical bet? If as has been classically held, democracy is contingent upon an enlightened populace, it stands to reason that it would also be possible for a population to lose its capacity for self-governance. Maybe Plato's cyclical view of how political forms transition and cycle through one another, where because of its internal tendencies democracy gives way to tyranny, is uncomfortably close to correct.

If not, how might things be retooled in order to achieve a higher level of civic sagacity, if such a thing is possible at all? It might also be wondered whether the present scalable one-size-fits-all "factory" system should be disbanded as we fragment unsalvageably into our various identity tribes? Perhaps policy reforms facilitating a devolution in federal and state control might contribute to a new localism (or even a new identity-inflected tribalism) that would enable stakeholders to draw closer to nearby educational projects that are more ideologically congenial to them and their overall vision of what is worthwhile in life. As in what Hegel idealized as the "folk religion" of ancient Greece and medieval Christendom[25]—and one can imagine an infinite variety of contemporary *Weltanschauung*—educational endeavors could be freed to move more decisively toward an integration with what John Rawls called various "comprehensive conceptions of the good" (i.e., conceptions of "what is of value in human life") rather than the "thin" procedural norms that are necessitated by the heterogeneity and gigantism of current systems.[26] The ethical problems with such imagined heterogeneity are of course legion, and one worries about federally guaranteed universal principles of inclusion such as various forms of nondiscrimination (e.g., racial, gender, disability, etc.). On this model there remains no guarantee of enforcing universal principles on the constitutional model of civil rights intervention; it may matter far more just *where* one lives (e.g., Louisiana vs. Vermont), the ante-upped situation that is reflective of the moral contingency built into localism.

But before anyone can rethink anything it is first necessary to consider those who are to do the rethinking. What might cause them to engage in such an enterprise? What might motivate a philosophical activity like rethinking basic assumptions that is for normal people a painful detour from everyday life and what really matters? Those afflicted with philosophy can't help themselves, but surely, with apologies to Aristotle and others, the philosophical *vita contemplativa* is very far from a popular comprehensive conception of the good, at least in the sense of offering a total vision of why life is worthwhile.[27] It leaves out far too much for most of us. Most people tap their reserves of deep philosophical thinking only when confronting some obstacle or crisis that demands it. Education's legitimation crisis may be such an occasion. For the legitimation crisis arising among educational (and other) institutions because of the twilight of certain economic gods is not far off in some distant future, but it has already begun. But its incipience is mostly indirect and opaque and in my view it rises around us in forms that are not so readily recognizable as frontal challenges to key institutional assumptions. It's not that obvious.

I contend, for example, that the welcome phenomenon of increasing acceptance of gender and other kinds of identity fluidity among the rising generation is—whatever else it is—a deep symptom of a loss of economic faith. Consequently, it might generate a rethinking about the approved paths in life, paths that are premised on the assumption that existing occupational niches are forever stable. It is not simply that the proliferation of identities is a result of moral progress or some preordained arc of history, but rather that they are rushing in to fill a human void in the society that has been left to us by neoliberal capitalism, where our occupations and therefore connectedness to society have crossed a threshold of precariousness, a liquidity threshold to paraphrase sociologist Zygmunt Bauman.[28] As per the old Marxist thesis about the material bases of all ideologies, what I am calling identity proliferation is not occurring simply by chance and neither is it on its own singular moral track somehow parallel to but disconnected from economic realities. While not reducible to them in any simple sense, our ideas are still ultimately rooted in our material conditions.

My basic thesis is that identity proliferation is tied to large-scale changes in the economy that have opened fissures from within which the identities emanate. These emanations necessitate a grand rethink of education broadly construed because they cannot be contained within the procrustean identity beds of past generations. The economy-education nexus that has come to rule childhood and adolescence—and stretches further qua "lifelong learning" and what philosopher Matthew Charles has termed the "pedagogization" of the life course—is at the very moment of its triumph dissolving before us.[29] As a result, as our eyes start to clear of those previous governing visions, as this twilight of our institutional idols deepens, we stand to undergo great alterations in self-understanding: the stories we tell ourselves about ourselves, our projects of self-narration, our identities, my sense of how I might connect to "the starry heavens above me and the moral law within me," as Kant elegantly summarizes. For the one constant is that we are fated to keep on narrating these grander visions to ourselves. One cannot help but attempt to continue doing so, to "see them before me and connect them immediately with the consciousness of my existence."[30] For better or worse, it is simply what we do as human beings, just as surely as when cold we seek warmth by the fire.

Introduction

*By 2050 a new class of people might emerge—the useless class.
People who are not just unemployed but unemployable.*
Yuval Noah Harari (2017)[1]

Once upon a time when labor demands were reliably high, the pre-eminent moral concern with capitalism had to do with exploitation. Liberal arguments about humaneness and excessiveness were deployed around important matters such as workplace safety, prohibiting child labor and limiting daily and weekly hours. But the most frontal arguments against exploitation were usually framed in Marxist terms as a campaign against the extraction of surplus value from workers' labor. Just as property itself is theft in some formulations, under capitalism labor is taken away from its rightful owner and illicitly transferred to an illegitimate appropriator.[2] The worker is being stolen from, *used*, her life sucked out of her—often literally in many industries—and into an ever-accumulating pool of quasi-autonomous capital. As Marx famously writes: "Capital is dead labour that, vampire-like, only lives by sucking living labour, and lives the more, the more labour it sucks. The time during which the labourer works, is the time during which the capitalist consumes the labour-power he has purchased of him. If the labourer consumes his disposable time for himself, he robs the capitalist."[3]

Stripped down to its barest moral elements, the implicit indignation at exploitation that runs throughout the Marxist canon is captured in the classic Kantian formulation that provides, in essence, a prohibition against unjustly *using* someone merely as a means to an end, reducing them wholly to a tool or thing; and, on a spectrum toward slavery, wage labor is conceived as dehumanizing in this manner. Accordingly, whereas the liberal left has tried to mitigate exploitation through reform, the Marxist left traditionally sought to overturn and end this kind of exploitation altogether as in principle immoral.

A major premise of the present work is that the underlying material conditions that made this left-liberal Kantian critical-moral paradigm relevant are shifting beneath our feet. Rapidly. Due to unprecedented levels of automation and associated phenomena such as globalization and outsourcing, people in the MEDCs are beginning on a wide-scale to face a "post-productivist" situation in which there is a severe reduction in the need for even the exploitation of their labor. Many would welcome the chance to be exploited economically, in other words they would welcome a living-wage job, *any* job. I contend that severe consequences are unfolding from these developments, many of which lie far beyond the bare unemployment statistics. For one thing, the culture and mindset built up in the MEDCs, during conditions of near full employment in the immediate postwar generations, placed vocation at the center of identity, given the gendered division of labor of that era, for males as wage earners and "providers" and for females as "housewives" and "mothers." Although they are often stretched thin by the dual roles and there are many holdover traditionalists, women in the postwar period have also nestled into the previously masculine vocational identities *en masse* such that the workplace anchoring dynamics have started to apply to many of them in nearly the same way. However, persisting disparities in domestic responsibilities, especially regarding multi-generational care giving, have preserved an uneven identity terrain. The anchoring processes in such cases can have altered dynamics, e.g., a single mother who knows full well "who" she is but for all that is in more desperate financial straits than she would be if partnered. Whatever their complexity, these occupational identities— whether domestic or workplace-based, or both—tend to be internalized as central to individuals' sense of self, among the primary "go-to" answers people give when asked *"who* are you?"

It is underappreciated how securely individual identities have been rooted in vocation along with, to be sure, their other anchors such as religion and extended family. (Any sensible social theorist is averse to monocausal explanations.) I call this rooting phenomenon "vocational anchoring." When, due to the abovementioned widespread diminution of workplace labor needs, those vocational anchors start to loosen, funny things start to happen with human identities. We are capable of thinking of ourselves in all sorts of ways but, given the material bases of social reproduction, the vocational anchoring is the one that tends to hold everything else together, durably to connect individuals and families to society. Zygmunt Bauman, whose analysis is foundational for my own, identifies this "changed nature of work" as central to what he calls "liquid modernity," where work is:

> cut out from the grand design of humankind's universally shared mission and no less grandiose design of a lifelong vocation. Stripped of its eschatological trappings and cut off from its metaphysical roots, work has lost the centrality which it was assigned in the galaxy of values dominant in the era of solid modernity and heavy capitalism. Work can no longer offer the secure axis around which to wrap and fix self-definitions, identities and life-projects.[4]

There are other vestigial social glues in this respect such as religion, but on balance in American and western European life religious belief and affiliation has severely weakened as an identity anchorage. When labor is chronically less needed — or equivalently when the nature of that labor is altered in certain ways — whole populations can experience Harvey's "global alienation": they start failing to find meaning in their jobs, their institutions, their communities and even their personal lives. These alienated populations are a key factor in the social and political instability that we now observe across the globe. Such groups are prominent within the geographical and often environmental dead zones not lifted by the updraft of globalism, for example, in depopulated and otherwise forgotten rural areas—and they often lash out in unpredictable and alarming ways.[5] Perhaps out of enlightened self-interest these individuals will support humane social welfarists (and moral universalists) like Bernie Sanders or Jeremy Corbyn. But they may just as easily choose a short-term catharsis and elect Donald Trump, vote Brexit or support rising ultranationalist political parties.

Ex hypothesi the perturbations of the not-so-satisfied also signal a profound moral shift that is capable of taking us beyond the Kantian moral paradigm. Previously, striving to be free from exploitation could also be thought of as a demand to be treated as an end in oneself and not merely as someone else's means. However, when exploitative conditions are not even available— when industrial capitalist *abuse* has become neoliberal capitalist *neglect*—the social demand to be treated as an end in oneself, commonly manifest as a demand for dignity and respect, remains constant, but logically it cannot be satisfied by non-exploitation—because, in defiance of traditional expectations, exploitation is precisely what is no longer occurring. Instead, because the basic need for it persists, human dignity is sought elsewhere, primarily in alternative respect-garnering loci to be found among the proliferation of identities other than those that are job-based. One *re-anchors*, that is, one ties oneself back into humanity, via the alternative identities that are actually available including sometimes religious revivalism (witness the resurgence of charismatic and evangelical Christianity in both MEDCs and LEDCs). Or—and in some ways this is the interesting part—one adapts or invents new ones. Not just "alt-right" or "alt-left," more and more of us are becoming "alt-something"; and as we are automated into uselessness, we are also alt-ed into a new and vastly more variegated identity terrain where meaning-seeking is redirected. We are now seeing the beginnings of a world historical spray pattern of new and previously unimagined forms of human self-understanding. And the sponge-wielding worried moralists will be unable to soak it all up; it is too big of a mess and it grows faster than the absorptive capacities of the sponges. This fluid moment holds promise but it also carries with it great dangers.

This book has four aims: (1) zero in on the underlying moral dynamics of identity proliferation; (2) examine some of the effects of the de- and re-anchoring process, especially in areas such as education where these effects may be expected to be acute; (3) assess the possible human responses to the challenges; (4) defend the outlines of what I see as a preferred alternative among them. This endeavor is primarily a *synthetic* inquiry in the sense that it aims to place together apparently disparate happenings and reveal their interconnections. These are phenomena that *appear* in the daily news in all sorts of guises— transgender rights, Make America Great Again, NFL protests, UC Berkeley provocateurs, forest tree sitters, #metoo, Facebook gender options, Twitter, Tumblr and 4chan alt-groups, kinky lifestyles, fundamentalist homeschoolers, neo-Stoic and neo-pagan revivalists, *ad infinitum*—whose very quantity, ubiquity and intensity is *itself* suggestive and undertheorized.

Much more is going on with identity than the cartoonish "identity politics" of street protests, Hollywood and college campuses. My suspicion is that these are all just symptoms of a malady we have still not quite put our finger on. "Neoliberalism" or "capitalism" *simpliciter* are tempting identifiers, but even these labels fall short because they tend to be too narrowly economistic and one-dimensional. Despite their ubiquity—we have now grown legion in academia—the critics of neoliberalism tend to assume as self-evident the sources of their moral outrage rather than argue for their legitimacy. My project is to get closer to these sources and examine, in the simplest possible terms, the basic moral paradigm within which they operate. Perhaps because of the herd dynamics inherent in political life there has been surprisingly little self-examination by the left of its motivational sources. Psychological aspects of all this have been recently explored in interesting ways by Joshua Greene, Jonathan Haidt and others.[6] It is also a purely philosophical inquiry in the Socratic sense of trying to examine one's deepest motivations and their defensibility.

Chapter 1 explores something off the radar of economists' analyses of un- and under-employment: the threat that a dearth of stable, meaningful jobs poses to individuals' sense of self-worth and social connectedness. The *locus classicus* for this general concern is still found in Kant's "formula of humanity" in which we are obliged to treat one another as more than mere means to our own predetermined ends. This basic deontological principle is canonical in philosophy and is perhaps not so exciting to contemporary theorists because it smacks of the introductory college course and provokes a jaded "been there done that" attitude. Yet this foundational notion of respect that Kant emphasizes still provides the boiled down moral core of many of the more fashionable moral causes, including the various permutations of identity politics and intersectionality. Central to our sense of own respect worthiness are the stories we tell about ourselves to ourselves, our propensity for self-narration, the most minimal level of which is a basic gestalt whereby we see ourselves as a "fit" with something—anything—larger than we are in isolation. For many during the postwar economic boom, a solid and steady job was an integral aspect of that connecting mechanism where the answer to the question of *who I am and how I'm part of a "we"* was answered in large part by my occupation. This ability to place oneself into the larger social framework now seems threatened on a large scale, not only through the lack of steady jobs but by a draining of meaning from those that remain. The threat itself in turn makes clear what some would term an uncomfortably conservative flip side of the respect demand, namely, an obligation toward *respectability*, a notion that should be rehabilitated from its Victorian overtones.

It is not only a problem of employment and income but also a psychological and even philosophical problem having to do with our ability to find meaning—even when we do have jobs—in a workplace that seems for so many to be void of larger purpose. This *existential* vocational vacuum is coming to be filled by a Pandora's Box of identity configurations, some old, some borrowed and some seemingly novel. In disciplines as diverse as international politics and clinical psychology, one hears mention of a "new tribalism" with unlimited affiliative possibilities.[7] If one cannot *place oneself* among others through work one does it in some other way, through aspects of oneself that one takes to be at one's core: not only the "official" social justice identities like race, class, gender and sexual orientation, but also a blooming variety of identity alternatives both within or outside the traditional categories. These burgeoning neo-tribal identities are starting to make greater sense to people than do the pre-fabricated fare that has been foisted upon them by identity mongers across the political spectrum. Whatever one "chooses," what remains the case is that, in an unexpected twist on Kant, in addition to seeing others in a certain way, one needs to see *oneself* in a certain way as well; one needs to *be* something too. My view is that this basic narrativizing need is far more powerful than the due it has thus far been given. Amidst the demise of the great religious metanarratives, it has yielded a veritable identity factory that reproduces traditional identity formations and churns out new ones at an increasing pace in order to fill the connective gap left in the wake of the neoliberal phase of capitalism, where we have passed from a period of relative occupational durability through which people could once reliably connect to society. Without that stable occupational anchor, and aided by the intensified global internet connectivity

23

now possible, a vastly increased range of identity options is now available to be seized upon. This incipient affiliative explosion is in turn breaking open the most basic structures and conceived purposes of previously stable-seeming norms and institutions.

Chapter 2 explores the challenges this liquefaction of identities poses to the institution historically tasked with molding and shaping them: education. The classical notion of what the historian Werner Jaeger called an orienting *paideia* recedes farther and farther from view: the self-conscious collective attempt to render the rising generation according to a determinate cultural ideal. Identity proliferation constitutes a direct challenge to the idea of public education as a common project dedicated to the reproduction of shared values. For without any kind of paideutic ideal, education is formally nihilistic; by definition it becomes axiologically directionless. It carries on with no regulative image of the type of person it wishes to foster. Performatively it believes in nothing and at best runs on the fumes of atavistic ideals. This is what we see unfolding currently, in a way that is particularly striking in higher education, which under neoliberalism must like everything else frantically market itself to consumers—or, in policy jargon, "stakeholders." The extreme heterogeneity of identity proliferation puts further pressure on an American school system already legally designed for content-neutrality concerning religion and other matters of ultimate aims and purposes. Extant constitutional neutralism and identity proliferation combine to open up a new front in education's legitimation crisis as the system can only appear increasingly nihilistic and unable to chart any particular course other than self-perpetuation through stakeholder satisfaction. It thus becomes indistinguishable from a business corporation dedicated to mere self-perpetuation and quantitative growth (i.e., profitability) and loses any unique claim to be serving the public good, an ideal that has come to be seen as a mere sentimental relic.

24

In this setting, educational institutions appear as if they stand for nothing but themselves and are thus open to question once they cease to provide the promised consumer-driven outcomes, viz., jobs for the paying customers and a large enough pool of sufficiently skilled workers to keep those jobs' wages down. When for macroeconomic reasons these two functions are no longer being fulfilled, conditions become ripe for a wholesale systemic de-scaling; the education bubble bursts in the manner of a financial crisis driven by overvalued and under-secured assets and educators are stranded without their economic *raison d'être*. What to do with these gigantic white elephant school systems then becomes a political matter.

Consequently, my **Chapter 3** turns to politics. Many feel that politics provides a defensible vantage point from which to pursue a better vision of society. The left proffers at its core an egalitarian vision that is grounded in a conception of universal respect for one's fellow human beings. Yet both liberals and Marxists have long resisted any deep account of their own normative underpinnings (anarchists have done better), typically preferring a crypto-normativism in which moral commitments are either denied outright, as under allegedly "scientific" socialism, or via avoidance devolving into a raw emotivism in which the correctness of a position is based upon how strongly it is *felt*, usually via a public display of moral superiority; "this little light of mine, I'm gonna let it shine," as the old protest song goes. (Note the persistence of the demonstration as the left's favored political tactic and how comparatively rare it has been for the right, which has long favored different *modi operandi*.) The former stance of claiming to be offering objective scientific description has been out of favor since not long after its nineteenth-century debut, its most famous proponent probably being Friedrich Engels, whose candid goal was "to make a science of socialism."[8]

The remaining normative self-conception, emotive moral

25

superiority, has in my view severe shortcomings. Despite its self-assurance, it is not necessarily tied to any particular worldview or comprehensive conception of the good. Unlike the right, which has an easier time basing its precepts on orthodoxies of various sorts due to the inertial power of traditionalism, this untetheredness causes the left to devolve into a thin proceduralism that is most comfortable with legalism and a kind of juridical mode for social remediation.[9] This strategy is well-suited to winning battles against particular injustices, the marquee example being the civil rights struggle against legal Jim Crow *de jure* racial discrimination—understanding that this struggle was importantly motivated and maintained by the vestigial religious traditions of the black Church.[10] But it is also well-suited to losing the wars in the long run. This is because the proceduralism of legislative and legal reform runs out of steam over time as the norms that once animated it erode into an oblivion of bureaucratic regularism; civil rights becomes less a moral awakening than an interminable series of lawsuits.

On a larger scale, this is also the case with the theism and notions of human progress animating federal constitutional norms such as equal protection, non-establishment of religion, free speech and the like. As with Martin Luther King Jr's or the later Malcolm X's theological motivations, the worldviews responsible for originating the left-liberal agenda are eventually dismissed almost as embarrassments by the modern savvy and ironic progressive who no longer needs her grandparents' outdated beliefs. This is fine until it is realized how the eponymously labeled "progressives" must as the label suggests maintain a faith in "progress" but without any ordinates upon which said progress can be mapped. How do we know where exactly we are positioned on the map? Without an orienting vision, how can we tell if our movements progress or regress? Are there criteria by which one could tell? What would be implied by the existence of such criteria? Analogous to the situation with public schools, the absence of any shared conception of the good corrodes commitment over time. However indefensible their substantive visions (e.g., biblical literalism, market, fundamentalism, ethnic chauvinism), in this respect the far right enjoys a structural advantage as the mere presence of a worldview —*any* worldview—tends to defeat its absence; a bad vision beats no vision every time. Fragmenting identity politics and its attendant online "call-out culture" only accentuates these dynamics.

There is only one viable long-term response: *world building.* But nobody wants to hear this answer, as it cuts too far against our strong tech and consumer bias toward presentism and immediate gratification. It would be much easier for some quick non-ideological fix "to restore the Republic" like a Russia or porn-star scandal. We short-attentioned contemporaries find it very hard to think in terms of perdurable outcomes that may only be achievable after we are dead. King's searing final speech, "I've Been to the Mountaintop," is justly famous but the depth of the moral challenge it represents is rarely appreciated. In the 1968 speech, delivered in support of striking sanitation workers in Memphis, the ones who picketed with their iconic "I am a Man" placards, King articulates the religious basis for his political commitments. His memorable finale describes how political commitment must be based upon something larger that will outlive one's personal existence:

> Well, I don't know what will happen now. We've got some difficult days ahead. But it really doesn't matter with me now, because I've been to the mountaintop. And I don't mind. Like anybody, I would like to live - a long life; longevity has its place. But I'm not concerned about that now. I just want to do God's will. And He's allowed me to go up to the mountain. And I've looked over. And I've seen the Promised Land. I may not get there with you. But I want you to know tonight, that

we, as a people, will get to the Promised Land. So I'm happy, tonight. I'm not worried about anything. I'm not fearing any man. *Mine eyes have seen the glory of the coming of the Lord.*[11]

Most discussion of this passage understandably focuses on its eerie presentiment of King's murder the following day. But the passage's substance is memorable too. King indicates that his personal efforts must be placed into a larger, even cosmic context for them to make sense and, implicitly, to give them lasting meaning. It may be extrapolated from this that if one lacks such a cosmic context, some larger way to frame one's efforts that speaks to final purposes, some conception of the promised land to which the struggle and sacrifice is meant to lead, then one's individual efforts *do* in fact lack meaning and are *not* worth the giving over of one's life. As King had also famously said (1963), "if a man has not discovered something that he will die for, he isn't fit to live."[12]

King's final speech in Memphis reminds us that political activism cannot simply be about decrying injustice, no matter how urgent and just the cause. It requires a task that is just as difficult in its own way, the enterprise of building as a solid platform for the critique a meaningfully rich and inviting view of life that is answerable, in effect, to the grand Kantian questions of God, freedom and immortality interpreted broadly, in short, a view of life that speaks to who we are and what are our final purposes. If this human need is sufficiently appreciated, it turns out that, interestingly, the most urgent political task is actually non-political (or extra-political); it is more a grand philosophical and aesthetic enterprise than it is anything to do with organizing techniques or communications strategies. It must also be sincerely held, a requirement that is bound to be tricky because it is therefore unlikely to find total accord with *any* political ideology.

One cannot of course just foist some new worldview on people overnight. But the history of political and religious movements is replete with examples of small seemingly crazy sects in time gaining wide adherence, e.g., the *avant la lettre* transhumanist movements known as Christianity and Bolshevism, both of which projected an entire comprehensive conception of the good onto the widest screens of humanity. However tiny their provenance, such movements in retrospect had an ability to *resonate* historically within their parent traditions and also psychologically with the individuals they helped form.

The left should avoid altogether the fork in the road leading to its two dead-end tendencies of *modus vivendi* proceduralism // and identity authoritarianism. It must instead go *off-road* into the most aleatory of ideological projects in order to locate and construct a comprehensive and sustaining worldview capable of premising its defining egalitarianism. This ultimately is the educational project that people who are becoming useless — that is, *all* of us — need most of all. Yet it is an enterprise that must be undertaken with a minimum of *a priori* expectations, otherwise it would be merely an elaborate marketing exercise; if it is to be earnestly-held, a holistic conception of the meaning of life cannot be subject to political censorship. This is why such a quixotic undertaking has mostly been avoided by contemporary political types, especially in a heterogeneous society where it is not possible to make a common religious appeal for, say, a "Christian socialism" or suchlike. The depth-psychological or, if you like, "spiritual" examination that is needed is inherently contingent and unpredictable; one might end up believing different things coming out of it than one believed going in. Along the way, a sincerely philosophical temperament would seem helpful, that is, an ability to change one's mind in light of what seems true.

In **Chapter 4** I begin to address this set of challenges by sketching the primary worldview alternatives that seem plausibly on offer, not in principle but *realistically* given our present historical and cultural context. Then in **Chapter 5** I make the general case for an approach that seems to me most fitting, one that will be recognizable but also alien and perhaps even threatening in several respects. As cynicism is illegitimate for this enterprise, and by "cynicism" here I mean the advocacy of a worldview on a utilitarian basis merely for its political *outcomes*, I will own up to being a true believer first and a strategist only later. I will not pretend to the role of disinterested scholarly narrator but rather I will advocate—hopefully in a responsible manner—for a particular direction. The paradox here is that a defensible politics can only be advanced by a sincere and therefore uncertain detour via the extra-political; one cannot "cheat" by insisting on predetermined political desiderata. When you are at the cliff's edge and what stands behind you is immovable and unacceptable, a leap forward into the dark is the option that remains. Though fraught and dangerous, it is a comparatively survival-positive move because at least it gives you a fighting chance; in other words, it's so crazy that. . .*it just might work.*

Chapter 1

Identity proliferation

Men and women look for groups to which they can belong, certainly and forever, in a world in which all else is moving and shifting, in which nothing else is certain.
Eric Hobsbawm (1998)[1]

Respectable creatures: From autonomy to automation

Kant's "formula of humanity" enjoins rational beings to treat one another never merely as means but always also as ends in themselves: "Act in such a way that you treat humanity, whether in your own person or in any other person, always at the same time as an end, never merely as a means."[2] Correlatively, human beings command dignity, respect and autonomy, aspects of a core Enlightenment moral attitude which, however much it falls short in practice, is associated with hallmark liberal political norms such as equal treatment under the law, civil liberties, religious tolerance, non-exploitation and personal autonomy—the latter in areas as varied as limits on surveillance and reproductive rights. The old Marxist dialectical view is that these "bourgeois freedoms," while welcome and necessary, were fated eventually to be superseded by allegedly superior post-capitalist forms of "positive freedom" arising from more evolved norms of production and distribution.

These successor norms were never quite specified, however. As per Marxist materialism, the implied idea is that the animating ideals of human interrelation would take care of themselves; phenomena such as alienation and wage exploitation would diminish to a vanishing point as a function of salutary alterations in real economic conditions, either by evolutionary or revolutionary means. Yet the normative underpinnings of the transition from exploitative wage relations were never really

31

adequately theorized. In cruder circles it was even set aside as a depoliticizing and hence distracting bourgeois concern. Yet it is famously, even notoriously, true that, despite its pretenses to objective science and denunciations of moralism, Marxism (including Engels) was driven by an animus toward what it perceived as injustice and human suffering. And it does not seem hugely mysterious to find in their denunciations of the "theft" of workers' surplus labor and the desiccating exploitation of the "vampire" Capital a tacit allegiance to the broad Kantian formula of humanity and its injunction against treating people as mere means. Why else would one be bothered by workers' sorry fates?

Put another way, even if one viewed socialist revolution as inevitable, why *welcome* it rather than merely be *resigned* to it? Whence the approbative attitude? Welcoming it implies a positive valuation, a sense that it represents a state of affairs more desirable than the one that had preceded it. And presumably it is more desirable not because of metaphysics (Marx's starting point is the rejection of an overly abstract Hegelian "idealism" after all) but because it is expected to make life *better* for the majority, where "better" means less exploitative and so closer to the Kantian "kingdom of ends" where social relations are constructed such that we honor one another's humanity in some yet-to-be-determined manner. Despite the rhetoric, there is no "overcoming" of boring old Kantianism at the moral level. The essential moral motivation for Marxist revolution still springs from an imperative to secure human autonomy. This is the only way to make sense of Marxism as the normative view it is advertized by Marx himself to be, i.e., where instead of being like philosophers who have "only interpreted the world in various ways; the point is to change it."[3] It is hard to make sense of that famous finale to the *Theses on Feuerbach* in any other way than as implying a normative "ought." Social relations and material conditions *ought* to be harmonized with the real needs of humanity.

However, the revolution never occurred in the leading MEDCs as dialectical materialism had envisioned. (The idea was that the productive forces had to be developed to a certain degree for the exploited classes to have "nothing to lose but their chains" and gain sufficient self-consciousness and thus revolutionary agency.) In the leading economies (England, France, Germany, the US), the revolution either neither came or it was defeated or deflected into a welfare liberalism that was a poor and painfully slow substitute for even "evolutionary socialism." Because of this, some Marxists have made the mistake of assuming that the forces of reaction were passive and static and could only change via revolutionary opposition, to think that change to the system must always come from the morally pure rebels on the outside looking in. But this is to ignore one of the deepest tenets of the dialectical view of history, viz., that every system contains internal tensions that drive it to alter itself over time. This is as true of the nineteenth-century counter-revolutionary settlements as it is of anything else.

Among the most powerful of these tensions was that of economic competition leading to massive and literally earth-shaking technological developments that, in turn, further accelerated economic competition. Labor became vastly more productive, having now at its behest augmentations in energy (e.g., wind to steam to coal to fossil fuels) and communication (letters to telegraph to wireless to radio and television to the internet) and many other areas. And over time and in the long run, as labor becomes more productive, less individual producers of it are needed, less "variable" or "human capital" in the Marxist lingo. In fits and starts, processes of automation begin to develop, sometimes in ways that seemed to demand *more* cheap labor (the cotton gin, telephone switchboards, factory outsourcing) but eventually via ceaseless technological innovation and development, automation began to outpace and cannibalize the new labor needs it created and thus started to characterize production across the board. We are at the point where it would be foolish to bet against the automation of any particular occupation, as previous candidates proposed for protected status seem to fall by the wayside daily, including white-collar jobs once thought immune such as insurance adjustors, radiologists, lawyers, journalists and professors, to name a few. The dire warnings reflected in titles and headlines like these are becoming ubiquitous and reflect this growing concern: "Rise of the Robots: Technology and the Threat of a Future"; "Why the Future is Workless"; "Humans Need Not Apply" and "Robots will take our jobs. We'd better plan now, before it's too late."[4] Tech entrepreneur Andrew Yang forthrightly declares in *The War on Normal People* (2018), "I am writing from inside the tech bubble to let you know that we are coming for your jobs."[5]

It turns out that the development of productive forces has given rise to a metastasizing automation that threatens all-too literally to fulfill Enlightenment promises of freedom, tolerance and non-exploitation as more and more human beings are cut loose from labor and the productive processes and are thus "freed" from even the possibility of exploitation. Obscenely large and growing human masses would at this point feel genuinely fortunate to find someone to exploit them in order to earn a reliable living. Refugee hordes from Syraquistan, migrant masses from Mexico, along with climate, economic and other political refugees with nowhere to flee—including the vast ranks of the incapacitated and incarcerated—daily dramatize the concrete meaning of surplus humanity, those who have become "the disposables" among us, the "wasted lives."[6] And even for the fortunate ones, psychological miasmata such as addiction, anxiety, depression and suicidality debilitate bodies to the point of inoperability.

In a cruel irony, these are among the forms through which the masses are tragically "freed" from capitalist exploitation and placed outside the circulatory loop of surplus labor value. This is where the Kantian formula of humanity, that liberal centerpiece of moral confidence, in the economic arena now crumbles beneath our feet. As per the classic formulation, the worst economic outcome it envisions is radical exploitation, in the sense of being reduced to a mere means, an other-directed tool in someone else's workshop. Today's outcast humanity, however, show us that there is an even worse possibility waiting for us: precisely the status of non-exploitability, a vertiginous uselessness whose implications threaten everything to which we have become accustomed. Absent a concomitant Nietzschean "revaluation of all values,"[7] we are left with the stark moral-existential problem of how we are to relate to outcast and "useless" people, especially given that *all* of us seem to be falling down the same sinkhole of outsourced replaceability.

There are two broad avenues of response. One is the dark road, the left-hand path of nihilism and what I have previously termed "eliminationism."[8] Cull the herd. Like the horrifying Nazi campaign against the disabled (who were labeled "useless eaters"), those who no longer "contribute" should simply be made to vanish by whatever means, subject to some manner of what Saskia Sassen terms "expulsions."[9] This is an invitation to join neoliberalism's death cult, the murder-suicide that it envisions by its very internal logic, first for those deemed economically useless and disposable and second for everyone else as environmental externalities render the planet uninhabitable. One might call this deep nihilism a new formula of *inhumanity* that is, perhaps, not so widely appealing when stated as such. The other road, though, involves a recapturing of something as-yet-unforeseen that would constitute a *truly* post-Kantian, post-exploitative framing of human subjectivity and mutual obligation. The *avant-garde* philosophical, artistic and educational project here involves the exploration and articulation of what these post-work, post-everything modes of human—and perhaps, to use a fashionable term of art, "posthuman"—intersubjectivity might possibly be like. It may even be possible to indulge in optimism here, as widespread fear and survival-positive biological aversion may well drive us toward this more life-affirming frame. We may be seeing the first salutary tremors of political instability in the developed economies and at their peripheries, from Occupy to the Arab Spring to the first real challenges—from right and left—to the static political establishments in the West: Corbyn, Sanders and, yes, Trump, Brexit and Fidesz (Hungary).

The ideational challenge lies in our being forced to think beyond the Kantian means-ends moral framework and to learn to attend to questions that we actually feel in our lived experience, recapturing some of the attitude that existed among the existentialists of the postwar period and making up for a generation of theorists lost to neoliberal "postmodern" word games and postures. As Nietzsche saw, the confrontation with nihilism is not to be wished away by cleverness and part of its horror is its durability and extension through time. The twilight of neoliberal idols is starting to reveal the outlines of that confrontation with a clarity that has not been seen since the violent dawn of the capitalist age itself. For inevitably this rising tide of placeless humanity is becoming conscious of itself; the very tools that led to its creation—the technologies of globalization and automation—are now also allowing, as if part of the curriculum of a master course on Hegelian dialectic, people more clearly to see and comprehend their rootlessness and contingency. There is nothing so jolting as a confrontation with one's own mortality, which is precisely where moving beyond the Kantian formula of humanity takes us.

To be clear: in the old framework as long as we were not in the moral stance of using or exploiting someone, viz., treating them merely as means, we were *ipso facto* treating them as ends, either in some positive sense where we are actually interpersonally involved with them, say, via bonds of love or community, or negatively in the libertarian *modus vivendi* sense of "live and let live," where persons are shown respect implicitly via the acknowledgment that they are capable of directing their own lives. The economic ground beneath our feet has shifted, though, such that "live and let live" has indeed transformed into "live and let die" (just as it does in the Paul McCartney song). In terms of the material conditions of production this means that when the labor-intensive "all hands on deck" phase of industrial capitalism shifts into the current technology-enriched automated phase of augmented productivity (less labor intensive), we have transitioned from a setting of economic *abuse* to economic *neglect*, where those in the latter category are now rendered disposably external to the shrinking loop of labor-utilizing production. So being economically exploited is not at all the worst social fate, as any of those rotting in previous centuries' dungeons could readily explain. As in ancient times, the worst social fate is to be expelled, a sentence that could be relied on to result in social and/or physical death (traditionally these two were one and the same). The status of being *cast out* has long been feared and in the Judeo-Christian tradition is often represented as the archetype of human misery (e.g., the fall of Lucifer, sin as separation from God, thrown into "outer darkness," etc.).[10] This is beyond seeing someone as an exploitable resource, as material for one's own enterprises. In a sense it is seeing them as *leftover* material, qua Marx's surplus army of labor, but in an important sense it goes beyond even this: not only do the people in this category not contribute but they are a net drain, a mere cost, that efficiency dictates must in the long run be *solved*, a final solution if you will. And yes, the 1930s Nazi campaign against the disabled, and the utilization of its logic in the ensuing Holocaust, starkly anticipates the darkness potentially awaiting beyond the Kantian framework.

The frightening path of mass human neglect would seem to require a form of psychopathy, in the sense of a severe compartmentalization of what David Hume calls "fellow-feeling" — "a very powerful principle in human nature"[11] — that is, a restriction of one's intersubjective sympathies to a rapidly narrowing circle of significant others. A certain degree of this shutting oneself off is of course normal and necessary for sanity. (There is a story about the philosopher Simone Weil weeping at the news of casualties from an earthquake in China. One is left wondering whether this is an example of exquisite moral sensitivity—as in Simone de Beauvoir's encomium of "a heart that beat across the world" or is it some variety of mental illness—or perhaps both.) At the least people need to retrain themselves *not* to see others in a certain way, a feat that certainly seems possible. One thinks of newcomers to a big city who are initially shocked and then later inured to homeless people on the streets or a Western visitor to, say, India or Egypt, coming after a time to "accept" levels of severe poverty to which she was previously unaccustomed to viewing up close. If Humean fellow-feeling is natural so too must be the converse ability to disregard the outsider. "Human nature" seems quite capable of taking us in either direction; it seems as ever to be frustratingly both rigid and malleable.

Worth noting is that what is left unsaid in the Kantian formulation is the precise nature of this "human being" or "person" or "man" or "rational being" (Kant uses all of the equivalent German terms) who is not to be treated merely as a means. There is an unstated assumption that this quality of human-ness or rational being-ness can be readily identified in the one standing before me who I am to treat in a properly end-like manner. Maybe we could augment Kant's account with a Turing test requirement of some sort—thinking along the lines of *Blade Runner* and HBO's *Westworld*, perhaps there is a rational being threshold that might in principle be reachable by a robot—but there is still something missing. This has to do with, I think, proximity, or at least how one perceives proximity, what one might call *moral proximity*: my sense of who are the human beings around me, who does in fact inhabit my perceptual field and consequently in effect who counts and who doesn't. There is no guarantee that I will actually *encounter*, face-to-face as it were, this other putatively rational being in my everyday lifeworld, even if we inhabit the same physical geography. In fact, physical geography seems less relevant than ever. My online interlocutor from Iceland may be far more present to me than the homeless man near the bus stop whom I've decided habitually to ignore or the elderly shut in on my block who I don't even know about. Or, as it turns out, other erstwhile humans (or, if you like, "former persons" a.k.a. zombies) who no longer have in the above sense a *socio-economic* place and so have been rendered un-encounterable. This invisibility may sometimes be geographical as in the case of the shut in or, say, a barbed-wired migrant camp, or it may be socio-political in the sense of someone who just does not travel in any of my social circles, someone I would only see (literally) in a public space as more or less part of the scenery.

I think that for a variety of reasons, most importantly economic and technological, we are tending to see people more and more as part of the scenery — stage settings for our own selfie-narcissisms — rather than as contextualized and fully-dimensioned co-inhabitants of our lifeworld. It is akin to that mindset we get into while driving in traffic, where I curse the car that cuts me off in a way that I would never curse a pedestrian who accidentally stumbled in front of me on the sidewalk. Of course it is always possible for that moral distance to be overcome and the more often this happens the better. I remember once on the road having an angry exchange with a fellow motorist where we pulled over and, after the initial awkwardness of actually finding ourselves face-to-face, surprisingly ended up having a decent friendly conversation; it was as if the spell of insularity wore off when we exited our cars to confront one another. Or when you are in a very functional relationship with someone like the cashier at a checkout aisle and, initially annoyed at the delay, you realize she is crying because she just received horrible news about a family member. At that point, one shifts immediately into "human" mode and the petty annoyance vanishes—unless one is some species of moral monster. But if the conditions for actually encountering the person have eroded, then this moral phase shift into human mode may not be able to take place. Sitting in a car in traffic and waiting at a checkout queue in a store are both situations where individuals are functionally removed from one another's moral proximity and attention, yet these are not situations in which the temporary absence of fellow-feeling is inherently difficult to reinstate.

As we structure in more and more of these situations of moral absence—in online environments for example—and they characterize a greater percentage of our interactions with others, they begin to shape our habits and then overall character more durably. Add to this the technological incursions into our attention economy where, back at that checkout line with the sobbing cashier, I might have my head buried in my smart phone and not even notice her distress, it becomes apparent how all the little obstacles add up. One might argue that simultaneously I am rendered more open to distant others' distress, let's say it's my sister texting me at that moment because she needs to talk, and there would seem to be some truth in this; there is *some* gain going back the other way, just like the invention of the telephone has long made it possible to *be there* for someone over long distances. But in the moral cost-benefit analysis, increasing callousness and inattention to the people actually around us is a very high price to pay for the connective conveniences.

On the analogy of these kinds of personal interactions, I think those who have been placed outside of the loop of production, those without that precious golden ticket of a steady job, are being placed further and further to the outer edge of our moral proximity such that not only are they economically precarious

40

they are *morally* precarious as well. What I mean by moral precarity is that if one adds just one more distancing factor (say, they live outside one's economically homogeneous subdivision or they wear the wrong clothes or, perhaps, they are of an unfamiliar ethnicity or religion) then it is that much easier to push them off the edge of one's moral universe. Thus they exit the collective moral radar screen as it were, just as has been the case historically with other group distancings along racial, gender, class, religion, disability, etc. lines. In the main traditional religious *Weltanschauungen*, there is an ideologically central *place* for "the poor," they constitute objects of charity, *mitzvot*, *zakat* and *daana*. However these almsgiving customs function in these ideologies, the unfortunate ones are strongly and directly *present* within them in a way in which they are simply not for neoliberal moderns for whom their very existence constitutes an irritating performative contradiction.

What seems different this time is the extent of the moral write-off of the unexploitables simultaneous with the growing sense that I *myself* could be next and if not next then somewhere in the not too distant future. Now that white-collar and even professional jobs are also threatened by automation and outsourcing, chronic economic anxiety and pessimism afflict just about everyone. Each of us—even those of us in what Stanley Aronowitz in 2001 called "the last good job in America" (the professoriate)—are just one or two technological developments away from being outsourced to a Bangalore call center or replaced by an online course. And because of their late-comer status and lack of social capital, the rising generation is even more afflicted with this permeating sense of socio-economic contingency. So running the tape forward it is easy to envision a situation in which this sense of socio-economic outsider status, this awareness of contingency becomes more and more widespread, where outsiders are steadily turned into outcasts and "deplorables" into disposables.

We reach a post-Kantian and post-productivist position because the traditional Kantian moral framework cannot accommodate these emerging moral dynamics without significant alteration. The Kantian injunction never to treat others merely as a means functions only when others are visible *qua others*, that is, as legitimate actually existing attractors of our moral regard; to respect you my apparatus of moral perception first has to be engaged such that I *notice* you and am drawn in by you as a fellow rational being. When you vanish from my gaze altogether there is no possibility of a relationship characterized by mutual regard, where we are related in the nexus of mutual obligation that philosopher Stephen Darwall characterizes as "the second person standpoint."[12] Neither is there any possibility, really, of *violating* the Kantian injunction against reductionist exploitation. If we lack a relationship in the first place, we are in neither a normatively good nor a bad relation; we are in no relation at all.

The philosophical problem is that for the formula of humanity to have any field of application it must first make a hidden *a priori* assumption of a pre-existing sociality where rational beings necessarily are mutually recognizing. It is a reasonable assumption: one of the great themes of philosophy from Kant himself and his beloved contemporary Rousseau to contemporary theorists of the self such as Sartre and Foucault, is that selfhood is socially constructed. More, it is an *artifact* of a prior sociality that provides the pre-existing materials out of which it is constructed. Though it may have the qualia/feeling of introspection and intimacy, our identity and self-definition always requires others. Kant himself makes the general point succinctly as "we judge ourselves happy or unhappy only by making comparison with others."[13] Sartre memorably extends this point by suggesting that not only are others necessary for our sense of self, our refracted self-image, it is impossible for us to have any sense of self whatsoever without them. In the punch line of his play *No Exit*, Sartre memorably suggests that because of this necessity, "hell is other people"; *pace* the alleged purity of Cartesian introspection, we have no unfettered access to ourselves except, frustratingly, through whatever funhouse mirror gazes the others in our socio-perceptual field happen to provide for us. Whatever qualities I feel are "me" — tall, kind, husband, cousin, neighbor, taxpayer, physician, Latino/a, etc. — are only available as social categories that have been pre-fabricated by others.

Like the onion that is onion layers all the way down, there is no essential "me-substance" underlying it all, no epicenter of a "soul" or "who-I-really-am" to which the layers all adhere. In fact there is, literally *nothing*, that is, no-thing, underlying it all. Paradoxically, for Sartre, it is precisely this nothing, *"le néant,"* that is the ultimate source of humanness and freedom, where "existence precedes essence" as the slogan goes.[14]

This influential thesis about selfhood has many ramifications. Among them is that the formula of humanity is only operative on the assumption of a pre-existing intersubjective connectedness. One way of looking at it is that the Kantian formulation needs to be understood as precisely that, namely as *intersubjective*, i.e., a two-way street. What I mean by this is that the moral force of the formula of humanity is traditionally understood as directed *toward* other rational beings who themselves serve as the implied addressees. It is a one-way injunction that is directed to those who are presumably in a position to *treat* their fellows in a certain way or other. In a word it is addressed to the *treat-ors* rather than the objects of the treatment, the *treat-ees*. Awareness of the hidden premise of intersubjectivity unearths this uni-directionality and also points toward a remedy: the formula of humanity can also be understood with the perspectives switched, from the point of view not only of the one about to do the treating but from the perspective of the one who is to be the object of the treatment as well. This reversibility of moral regard is in fact consistent with the internal logic of the Kantian position.

Most obviously, the inclusion of oneself is clearly stated at the beginning of the formula of humanity itself where one is to treat the humanity *"in oneself"* as an end in itself. And there is plenty of textual support for this perspectival shift's acceptability elsewhere. Analogous to Peter Singer's Benthamite "principle of equal consideration of interests," the foundational dictum of utilitarianism that "everybody [is] to count for one, nobody for more than one," logically there is no warrant for excluding *oneself* among the beings who are to be treated properly.[15] The need to include oneself is further underscored in Kant's discussion of the duty to develop one's own talents rather than allowing them to atrophy[16] and his insistence that "the humanity in our own person" requires us to "make ourselves worthy of humanity."[17] (This is why I am using the label "post-Kantian" rather than "non" or "anti-Kantian": these are extensions of Kantian themes rather than departures from them.) Anyway, what I am suggesting amounts to a kind of perspectival supplement in which, as Kant intended, one takes *oneself* into consideration as well, a frank and necessary circumscribed egoism that is justified by the assumption that one is *oneself* a member of the set of rational beings. A move like this is, incidentally, also implied in the notion of the "inalienability" of basic human rights as understood in the US Declaration of Independence; the enumerated rights must be understood as applying full force to *oneself* as to others to the extent that one cannot sever oneself from them even if one tried, e.g., one cannot sell oneself into slavery even if one chose to.[18]

This also overcomes a troubling paternalism inherent in discussions of "how we should treat others," especially where the objects of concern are "those less fortunate" and so on. There must be a complementary focus on the second person obligations that are owed to *us*; not only are there legitimate demands that others may make upon us, but also there are demands that we ought to make upon *them*. With regard to the Kantian formula, this means that I have as much of an obligation to demand respect *from* you as I have to proffer it *to* you. From a bit wider view, this means that I have a positive obligation to engage in and enact *respect-garnering* behaviors in order to guard against someone else devolving into treating me as a mere means. There is, in other words, a positive obligation to include myself with the rest of humanity by not allowing myself to be reduced to the status of anybody else's mere means; the prohibition against objectification applies to oneself as well. Liberation is as incumbent upon the un-liberated as it is upon anyone else. If this sounds like "blaming the victim," admittedly, it can be to an extent; my own oppression does not grant me an exemption from a duty to end oppression. It is obvious that there may exist severe constraints that limit the achievability of this ideal, but the obligation still persists, and when circumstances warrant, the Kantian view I am describing would counsel that there is a positive duty to rebel. This duty may be seen as analogous to the requirement in the US Military Code of Conduct for prisoners of war to attempt escape, and to aid fellow POWs in doing so where possible.[19] These are not blind injunctions to demand respect, rebel or escape in an unthinking manner, but general imperatives to work toward those states of affairs as conditions warrant.

So flipping the perspective of the formula of humanity makes apparent a reciprocal imperative to *secure* respect from others. One might say then that the hidden-in-plain-sight B-side of the Kantian imperative to respect others is that one must also make oneself *respectable*. The haute bourgeoisie of yesteryear would be disposed to hear this imperative toward respectability in terms of the social conventions associated with keeping up appearances and the like. But I would like to rehabilitate the notion of respectability from its latent Victorianism. While naturalism is always a risky venture, I take it as axiomatic that human beings typically, and maybe even naturally, crave recognition and respect, even though these elusive attitudes can be made manifest in an infinite variety of ways. Hegel, for example, in the *Phenomenology*'s famous "master-slave" dialectic, positions the interplay of mutual recognition as basic to all human sociality and politics.[20] So I mean "respectability" not in the narrow sense of middle-class propriety but in the vastly wider and more primal sense that flows from our nature as social creatures: to forge and maintain bonds of sociality and the concomitant actionable craving in us to win in whatever manner the basic regard of our fellow human beings. This does not mean their outright *approval* necessarily, as recognition can also come in the form of disapproval, but their basic recognition of us as one of them, one might say, as inhabitants of their second person standpoint ascriptions, their "you's" and/or their "we's." The minimal level of respectability would be the need to be seen *as* a human being by other human beings and correlatively the implementation of some means of securing that recognition. Human beings tend to deteriorate if they are isolated for too long and do not develop properly at all if that isolation occurs in childhood.[21] We are, as Aristotle famously holds, *zoon politikon*, animals for whom loneliness can be lethal.[22] This amounts to a basic need for minimal human recognition that may be characterized without too much fanfare as a human universal.[23]

The leading philosopher of political liberalism, John Rawls, goes even further, maintaining that "perhaps the most important primary good is that of self-respect."[24] Rawls recognizes the indispensability of what he calls self-respect: "It is clear then why self-respect is a primary good. Without it nothing may seem worth doing, or if some things have value for us, we lack the will to strive for them. All desire and activity becomes empty and vain and we sink into apathy and cynicism."[25] It is unsurprising to find, then, that the other side of the coin of the Kantian imperative turns out to be that, along with providing it to others, we are also obliged to seek that same respectability for *ourselves*. The two of course are interrelated. It is a commonplace of the world's wisdom traditions and pop psychology that one needs to have a certain self-regard if one is to be able to maintain a regard for others.[26] As Rawls put it, echoing such sentiments, "One who is confident in himself is not grudging in the appreciation of others."[27] The overall point here is to recognize with Rawls that this condition of self-respect or what I am calling respectability is not a superficial feel good notion or some other frill having to do with mere affect or "feelings," but is rather a structural condition of sociality, a bonding agent in the mortar that adheres individual and society. As Rawlsian philosopher Thomas Hill summarizes, "the crucial points are that self-respect is good or rational for anyone to want to have, that having it is a necessary psychological condition for enjoying other goods, and that whether or not members of a society have self-respect is significantly influenced by the basic structure of its institutions and its public rationale."[28] Importantly, it is a *reciprocal* relationship.

It is as if we cannot operate otherwise. For most people the pathways through which to garner respect are pre-fabricated; the winning strategy is to *conform* to certain expectations and *inhabit* certain social positions as ready-made bases for respectability. One's sense of self is powerfully shaped by these social roles and locations. I am a father, brother, Mayor, donor, taxpayer, neighbor, caregiver, pet rescuer and so on. For most people in recent generations, one's occupation in the economic sense provides one of the most powerful of these identity anchors; for so many of us (perhaps particularly Americans) we *are* what we *do* (for a living), this latter being inseparable from our sense of who we *really* are. In 2014 Gallup found that "US workers continue to be more likely to say they get their sense of identity from their job, 55%, as opposed to having their job just be something they do for a living, 42%. These results have been consistent

47

throughout multiple Gallup polls since 1989."[29] For college graduates, this number climbs to 70%.[30] Given this traditionally tight identification of identity and vocation, one of the hidden costs of the demise of occupational stability and durability for individuals is that, along with the income insecurity, geographical instability, psychological stress and other material consequences, there is often a very serious blow suffered by those individuals' identities. A career crisis therefore also often carries with it an identity crisis that, however widespread as in a financial crash, tends to be experienced individually as "I lost *my* job," etc. Some people may of course "land on their feet" with a new job or successfully transition to a new line of work. The change might even feel liberating such that one might come to embrace the ideology of the "lifelong learner" who is perpetually ready to upgrade skills, embrace mobility, "think outside the box" and so on.

At the other extreme there are those who have committed suicide due to their sense of self being so greatly wrapped up in a particular vocation-based identity that when it is suddenly removed they lose all of their bearings and cannot tolerate the ensuing socio-economic and psychological vertigo. (Worldwide, suicide rates surge as a function of unemployment and economic factors are estimated to be present in nearly 40% of suicides.)[31] Given the patriarchal and hetero-normative constitution of modern family arrangements, viz., husband and father as "provider," this situation is often intertwined with a specific crisis of masculinity among contemporary young men as those previously slotted for the "breadwinner" role must seek a substitute raison d'être.[32] A strong parallel may be seen among immigrant communities where individuals, especially former professionals, must find new lines of work because their former credentials are no longer recognized, for example, the Russian émigré music professor becomes a custodian in Peoria.

The larger point here is that because identities have been

so strongly psychologically anchored and socially integrated by *steady breadwinning-type jobs*, when these kinds of jobs dry up there are bound to be ramifications for those denied the accompanying social moorings. In Rawlsian language, these jobs have served as crucial social bases of individuals' self-esteem.

On the contemporary scene this socially integrative aspect of durable jobs is especially pertinent for the rising generation who in the US are less able to expect stable careers, to ever earn as much (especially factoring educational debt) as their parents, or at the same rates to experience many of the salutary effects of vocational anchoring such as geographical stability (including nearness to relatives) and home ownership, etc. As Bauman puts it, we are now in a situation where "secure jobs in secure companies seem to be the yarn of grandfathers' nostalgia."[33] If one's job as a factory worker or farmer or small shopkeeper or teacher or whatever provided a keystone as far as one's overall sense of self and social identity, the contemporary jobs crisis is much more than the quantitative statistic reported by economists as "unemployment" (which itself masks much of the larger picture, such as when in the US so-called "discouraged" workers, i.e., those not actively seeking employment over the last 6 months, are quietly dropped from unemployment denominators).[34] It is also an *existential* crisis for those individuals, their families and their communities, as they are so often then placed into disorienting and sometimes frightening psychological—or perhaps one should say philosophical—territory, that of having to find new answers to the question *"Who am I?"* And when this kind of question starts getting asked in a widespread enough manner, there is very little guarantee that the answers achieved by autonomous minds will accord with anyone's well-laid plans, particularly with the established and sanctioned identities of the past. There will be some, certainly, who will "rediscover" traditional identities, such as when a laid off banker spends more time with the kids and becomes perhaps for the first time a real father. Yet even this kind of thing would be an alteration of normative identities, because "stay at home father" was not really a respectable option in postwar America.

Even more interesting, though, is when those who are not incapacitated by their anchorless-ness, who are especially to be found among the young because so many of them have never known the vanishing job-anchored life in the first place, begin to discover or even perhaps *author* their own identity destinies. There ensues an ideological battle over these emerging identities and who gets to shape their nascent shapelessness. And there can be a dark menagerie of "winners" in this battle, those who construct seductive identities for the unsuspecting: cults, religious fanaticisms, racisms, fascisms, drug-based identities and many others. Given the wages of seduction—fame, power, money, sex—there will always be plenty of identity carnival barkers who may or may not even be true believers in the identity they're peddling.[35] One of the greatest admonitions of contemporary politics is never to forget the mass identity manipulations of the twentieth century, the ersatz "New Men" erected by totalitarian ideologues and "blood and soil" myth mongers. Those whom one might call mid-century's "identity floaters," including those displaced by war and economic depression, proved themselves to be spectacularly susceptible to these sorts of siren calls. If the last century has taught us anything, it is that populations who have been suddenly robbed of stable identity narratives constitute a social emergency in the making.

Just as it can be for individuals at the collective level, the inevitable re-anchoring process can become violent and all-consuming. On the contemporary scene, these are the people whom David Harvey argues suffer from a more thoroughgoing alienation concerning not just their jobs but their families, communities and their available political outlets.[36] Such populations are symptomatic of a growing political instability worldwide and they do not always appear as nice, peaceful, progressive protesters. They often appear as inscrutable and irrational destructive forces to their more "rational" occupationally-anchored compatriots; they do crazy things like support Trump or the European far right. On my analysis it is a mistake to view these impulses as mere revanchist forces wanting to turn back the progress of social justice (though there do exist revanchists); rather they are individuals trying to re-root their identities in violently destructive ways. And reactionary forces always have a built-in edge in appealing to the alienated as they possess a well-stocked cupboard of narratives of resentments and vestigial cultural prejudices that are nostalgically associated with better times.

It is necessary also to mention a further factor that gives greater power to the anchoring concern. The vocational meaning-drain is not at all limited to situations where jobs are outright lost via relocation or automation. Traditionally vocationally-anchored identities are not threatened merely by un- and under-employment. They are also threatened by alterations of the nature of the jobs that do remain. Anthropologist David Graeber in his widely-read essay "On the Phenomenon of Bullshit Jobs" memorably captures this concern.[37] In it Graeber tries to account for the fact that not only has all the automation and "labor-saving" machinery not resulted in a reduction in the quantity of work hours being performed, we seem to be experiencing a huge proliferation of service and administrative jobs that take place at farther and farther remove from the site of anything recognizable as production. Fewer industrial and agricultural workers and more financiers. Fewer professors and more higher education administrators. Fewer product developers and more advertising and media placement types. Fewer investigative journalists and more news aggregators and pundits. And so on, a trend that has involved "the creation of whole new industries like financial services or telemarketing, or the unprecedented expansion of sectors like corporate law, academic and health administration, human resources and public relations." Additionally, as Graeber points out, the growth in these areas does not even reflect "all those people whose job it is to provide administrative, technical or security support for these industries, or for that matter the whole host of ancillary industries (dog-washers, all-night pizza deliverymen) that only exist because everyone else is spending so much of their time working in all the other ones."[38]

This process has become familiar to everyone and is commonly reflected in an irony and cynicism regarding the workplace in popular culture, for example in such TV programs as the British and American versions of *The Office*, centering around a company that sells paper where it would be comical *within* the comedy for any of the characters actually to *care* about paper as such. One might call this the "Dunder Mifflinization" of the workplace (referencing the show's fictional paper company). As the protagonist office manager Michael Scott observes, "All of these jobs suck. I would rather live jobless, on a beach somewhere, off the money from a large inheritance, than have to work in any of these crapholes. They suck."[39]

As to what justifies labeling some jobs "bullshit jobs" rather than others, Graeber appeals to anecdotal evidence from his own experience where "after a few drinks" one's corporate interlocutor will spill concerning how meaningless his or her job is. I myself have had this same experience many times. Yet admittedly it is rather weak as evidence. I think Graeber should be cut some slack, though. Unlike unemployment statistics (which themselves are far less clear cut than it might seem and are fraught with ambiguity), nuances in how people conceive something as intimate as the meaningfulness of their jobs would seem quite difficult if not impossible to measure. Yet as always just because something cannot be measured accurately does not mean it does not exist. Abundant personal narratives to this effect and the fact that it seems to be a common theme across popular culture are suggestive data points to say the least. In any event, survey research may not be an appropriate mechanism for getting at this phenomenon. So it would seem open for philosophical speculation within the bounds of intuitive plausibility. Thus one may speculate that, given the growing complexity of production and distribution and their increasing distance from material production, contemporary workers often suffer from a kind of meaning deficit malaise. Meaning Deficit Disorder perhaps.

This meaning deficit is made even more acute in that it cuts into the teeth of an intensification of cultural expectations for *personal fulfillment* via one's job. This is a key point. It is not as if Graeber's bullshit jobs represents a new development that implies that past jobs were comparatively more fulfilling in and of themselves, that the work *itself*, say, on some stultifying factory assembly line doing mindlessly repetitive fine motor tasks, was somehow richer and more inherently satisfying than selling paper at the Dunder Mifflin office. Hardly. Valorizing the labor conditions and psychological effects of labor under heavy industrial capitalism would be to indulge in an egregiously inaccurate nostalgia; work sucked then too. But *ex hypothesi* it sucked in a different way. However ghastly and stunting the work-tasks themselves, under conditions of a more solid modernity it was much easier to conceive of the job in general "as the ethical axis of individual life."[40] The mere reality of boring repetitive tasks does not make for a bullshit job; occupational bullshitification is not a function of a job's *internal* qualities. Rather, it is a situation arising when, whatever its internal nature, a job is not integrated into any larger social whole and is experienced more purely as an *individualized* experience, which itself is symptomatic of market-myopic tendencies of a neoliberal economy that increasingly encourages all players to think of themselves as free agents dissociated from vestigial social bonds wherein, as Bauman puts it, "contemporary fears, anxieties and grievances are made to be suffered alone."[41]

In fact, corresponding to this macroeconomic individualization process, there is now a *greater* expectation that the internal qualities of one's job should be rewarding in the sense of personal fulfillment; ideally one's job should provide for "a purpose driven life," in the words of one bestselling spiritual guru.[42] Bauman identifies this aesthetic imperative as a core feature of liquid modernity, where work is to be as an "aesthetic experience":

> It is expected to be gratifying by and in itself rather than be measured by the genuine or putative effects it brings to one's brothers and sisters in humanity or to the might of the nation and country, let alone the bliss of future generations. ...It is instead measured and evaluated by its capacity to be entertaining and amusing, satisfying not so much the ethical, Promethean vocation of the producer and creator as the aesthetical needs and desires of the consumer, the seeker of sensations and collector of experiences.[43]

In this sense, one might have an entirely delightful and titillating bullshit job or, conversely, one whose day-to-day and hour-to-hour tasks are boring and repetitive (and perhaps dirty and dangerous) yet, because of its larger "fit" with one's perceived obligations and worldview, is sustainably full of purpose and meaning.

The vocational de-anchoring process is by no means complete, though; not all current jobs have yet succumbed to the abovementioned dynamics. Roughly three areas seem less prone to this meaning deficit (or aestheticization) problem:

(1) Jobs that directly involve helping other people, ideally face-to-face, where the beneficial results of one's labor can be observed in a reasonable length of time. Surgeons, nurses, physical and occupational therapists, teachers, social workers, EMT personnel, firefighters, horse trainers and addiction

counselors might be examples of this kind of work. Not that there can't be burnt-out and/or disconnected situations in these occupations, but reliably observing how one's work directly aids another actual person often provides powerful psychological sustenance.

(2) Occupations that, however menial, are tightly ideologically tied into the advancement of some believed-in worldview. So however repetitive and non-remunerative and often ambiguous, faith-suffused clerics or church custodians of whatever faith or, say, ideologically committed political party functionaries who might be doing boring repetitive jobs could well find the work so important because of the overall ideology it advances that they are able to persevere without deep cynicism.

(3) vestigial artisanal or neo-artisanal jobs where production processes from start to finish (or close to it) are executed by one person (or perhaps a small collective) who is thereby able to exhibit both creativity and technical mastery — cutting against the grain of the de-skilling process that has characterized capitalist material production since its inception.[44]

In short, tangibly and directly benefiting others' lives (not just selling to them, which involves a more strategic means-rather-than-ends stance), advancing a worldview one takes as worthwhile and personally conducting creative processes in the manner of an artisan would all seem to provide promising hedges against Graeberian bullshitification.

Yet the vast majority of jobs are not like this. This means that in addition to straight job loss and precarity, "post-work" also should be understood to include Graeber's bullshit or otherwise meaningless jobs, where "meaninglessness" is assigned by the jobs bearers themselves rather than some God's-eye view of which occupations are worthwhile and which are not. The question of identity is in large part a question of self-understanding after all, so the sense (or lack thereof) that one makes of one's own work is at least as relevant as data are to anyone's "objective" measure of value. It seems certain, though, that people can be mistaken on both ends: one might be delusionally convinced that one's hallway flower arrangements are of massive benefit to humanity or conversely a burnt-out surgeon may grow weary of saving yet another life. And this I suggest is an epistemological problem worthy of further philosophical attention; namely, the relation between objective reality (as they say), one's cognitive conception of it and one's psychological attitude toward it. My modest point here that the mere having of a job, even a relatively stable one, is no guaranteed protection against vocationally-based identity crises, in the manner of one of the manifestations of Harvey's global alienation. The state of not being able to identify with one's work comes in many forms.

But the social and psychological spaces cleared by the demise of durable and meaningful jobs also allows for a range of more interesting possibilities, a fluidity with regard to identity manifestation that would seem in principle to be limitless in its capacity to generate new and unanticipated forms. The hands and minds made redundant and idle through capitalism's relentless neoliberal phase become a devil's workshop of identity production, the power of which is quite rightly feared by anyone anywhere on the political spectrum—left, right or center—with an interest in pacifying the population, in counting and measuring them, manipulating their behavior, in short, rendering them *predictable*. Human nature abhors an identity vacuum yet precisely what new identities will be forged, along with whatever accompanying "festivals of atonement" and "sacred games" are to be invented alongside them is mostly impossible to predict.[45] One might make a few extrapolations based on current trends, e.g., more sexual and gender-based identities, but it is impossible to forecast too far ahead because there are simply too many material variables and too many peoples' minds involved. Add, say, pressures of climate change and/or war here or the adoption of new technologies there and add into the mix completely unforeseen variables and it becomes apparent that the only experts who may be "qualified" to speculate about the identity forms of the future are authors of speculative fiction, including those involved in the wildly popular apocalyptic/dystopian genres.

Liquid modernity and identity proliferation

All fixed, fast-frozen relations, with their train of ancient and venerable prejudices and opinions, are swept away, all new-formed ones become antiquated before they can ossify.
Karl Marx and Friedrich Engels (1848)[46]

To the rising generation, the expectation of vocational stability seems like an item of nostalgia, something to be viewed through the kodachrome palette of their grandparents' photo albums. Though in retrospect they may seem to have been born at the right time economically, those grandparents now have their own challenges. It is still hard to perceive as anything other than especially cruel the fate of the 61-year-old grandparent with health problems who is summarily pink-slipped out of her job and told to reinvent herself like a good "lifelong learner." The vestigial vocational anchoring of a person like that born in the 1950s is likely already to have shaped her basic outlook and although she can of course have any manner of new thoughts and gestalt shifts, she is in a different psychological and existential position from the twenty-something who has already seen through the promises. Both the sixty-something and the twenty-something are hybrid creatures with respect to a recent past where "who you are" was so commonly understood in terms of "what you do" and the emerging future that seems to be severing that identity nexus. But the twenty-something has typically been shoved farther along into the new territory by necessity, however sheltering or helicoptering or velcroing her parents might have been. She may have received participation medals at youth soccer matches, but she no longer sees her social participation, "taking her rightful place," as it were, in the simple and predictable terms of her parents and grandparents. Not that the previous middle-class generations had things easy, of course; they probably had them materially harder overall. But if they were able not to get ruined by war, the postwar American economic golden age provided *en masse* the psychological and epistemological comforts of knowing where one stood and what one had to do in order to be stable and successful. Now however the very goals of stability and success have become fluid and contestable such that it is hard even to begin to know what they might mean. This destabilization is an existential symptom of modernity liquefying beneath our very feet. And *ex hypothesi* this is only the beginning.

So how do people who are unable to answer that "who?" question in vocationally-anchored terms construct alternative bases of identity? Less directly connected to material production processes, at best inhabiting the far more ephemeral realm of perpetually shape-shifting service economy jobs, how do they understand themselves to be connected to other human beings and, above all, how do they understand *themselves*? Specifically, when identities are no longer related to the vocations that were once upon a time themselves reliably tied into the larger narratives of community, nation, religion and so on? It seems that we're not in Kansas anymore. So few of those now coming of age can anticipate durable jobs and careers that once anchored human identities both temporally in terms of the narrative arc of their lives and spatially in terms of their social place among others. With ready-made social identities to settle into thus becoming scarcer, how can we expect people to react to that unavailability?

⎰ ⟨⟩⟨⟩: How people describe the feeling of Trump's rallies.

One imagines what once seemed the limitless suburban tract housing developments all filling up and projecting a generalized "No Vacancy" message for the rising generation raised in them and then tossed out.

This is all very mixed news. On the one hand, it is devastating economically, as one needs somewhere to live and without ownership one is denied the great postwar American property values wealth escalator. But on the other hand one does get to escape the suburbs and its tyrannizing aesthetic. All the while, though, the placeless ones still need to inhabit *something*, *somewhere*; despite their relative economic placelessness, they do not, as of yet, simply vanish. (In the background, though, one still worries about the traditional "war trash"[47] solution to generational surplus and, more darkly, various other "human smoke"[48] scenarios that haunt our psyches.) As an implicit response to this situation, it is perhaps not accidental that "apocalyptic/dystopian" has emerged as perhaps the most popular book and film genre for the rising generation.[49]

The recent craze for dystopian aesthetics illustrates one aspect of how the anchorlessness and identity drift may be processed in popular imagination. Not in the particular resolutions to particular works of fiction—the plucky young heroine bringing down the system against all odds or the renegades limping through zombieland, etc.—but in the rise and popularity of the very genre itself. There is a certain psychological resiliency evident in this apocalypticism. If anthropology and the history of culture and religion show us anything, it is that we are irrepressible narrative engines. A cultural universal is clearly that human beings, wherever and however they find themselves, incessantly spin narratives about who they are and their place in the universe. We inevitably inhabit what communications scholar Walter Fisher calls a "narrative paradigm" that powerfully shapes most and maybe even all language and communication.[50] French literary theorist Roland Barthes elaborates this universality that is manifest in "almost infinite diversity of forms":

> narrative is present in every age, in every place, in every
> society; it begins with the very history of mankind and there
> nowhere is nor has been a people without narrative. All
> classes, all human groups, have their narratives, enjoyment
> of which is very often shared by men with different, even
> opposing, cultural backgrounds. Caring nothing for the
> division between good and bad literature, narrative is
> international, transhistorical, transcultural: it is simply there,
> like life itself.[51]

Forgive the cliché, but whatever else we are, we appear hard-wired incessantly to spin narratives—it appears to be one of those human universals. This tendency will undoubtedly continue among those who are uprooted or never rooted in the first place. In fact, it seems to me that we can expect a blossoming of creative efforts at narrating and re-narrating old, new and borrowed identities among those who are trying to root themselves in whatever ways are still available to them; disruptions can be aesthetically quite generative. The best artistic periods can be altogether miserable politically and economically, perhaps the best case in point being that the chronic instability of quattrocento Italy provided the setting for the Renaissance. Living in a time of great art may be bad news indeed.

As those living through it all become newly conscious of their geographical and moral placelessness, the narrative fabricator that is the human mind must once again construct forms of credulity upon which to endure and to procreate. A turn back toward the left-hand path of elimination must be rendered in a practical way unimaginable, just as suicide is experientially unimaginable to the mentally healthy. At present, as we see it unfold before us, on its own terms, the emerging post-neoliberal world beyond Kantian exploitation seems, like the fiction genre, to be more likely dystopian and apocalyptic than utopian. But looking more widely than is possible exclusively within the economic lens reveals greater complexity, particularly in other ideologically basic areas such as religion, sexuality and cultural affiliation. To paraphrase Marx, all that is solid seems to melt into air, not only in terms of the external realities of production but also with regard to the internal experiences of selfhood. This perception of fluidity is further symptomatic of Bauman's liquid modernity. Here, material policies of privatization and consumer-style individualism combine with widespread skepticism about previous generations' metanarratives to cause us to *internalize* a certain ambivalence, one that we *feel* in ways that are hard to pin down; we feel increasingly free floating or "liquid" vis-à-vis the more "solid" and durable forms that identity took previously. Sometimes we are happy to reject them but, often tragically, sometimes we feel we can "never live up to them." Working mothers may experience this dynamic as a "damned if you do or damned if you don't" dilemma as they try to balance career and family, for example.

Our self-definitions are experienced as increasingly unstable in at least two major respects: (1) there is a frenetic temporality, a "speed up" that is associated with the fluctuations of consumer choice as we try to "keep up" with everything we "need," and (2) we witness the bubbling up of proto-identities, newly recognized forms and combinations—such as Facebook's 70+ gender options—that seem perpetually to be generating themselves out of our surrounding social milieu. And many of these latter are continually transforming themselves into fully-fledged "identities" with their own moral and legal agendas (and concomitant opprobrium directed at those not nimble enough to recognize them at their inception, e.g., the labyrinthine neologisms perpetually emerging regarding gender identities). So there is an intensification that has both temporal and spatial dimensions. Less tethered to the anchoring vocational identities of yesteryear, we float freely into an acceleration process that tries perpetually to keep up with the shiny and the new as we see all around us—and sometimes within ourselves—the emergence of a cornucopia of new—or newly revealed—identity options, many of which were neither available nor even envisaged in previous eras.

Yet the phenomenon of identity proliferation is anciently rooted in the Western tradition, in both the religious (Jerusalem) and philosophical (Athens) traditions. Within those traditions, the Kantian conception is perhaps the apotheosis of an Enlightenment conception where human identity is conceived ultimately in terms of our capacity for autonomy; we are truly *ourselves* to the extent that we can exercise *our own* will and make *our own* choices. True morality for Kant involves a "self-legislation" where we do not blindly follow conventional rules but rather we *choose* to follow them because we understand and embrace them by our own lights. We are respect-worthy insofar as we are able to accomplish this kind of self-definition and our capacity to do so is the basis of our common humanity.

A difficulty arising from this picture is that however equal among one another we may be in *capacities*—which in legal and political terms are represented as our being the bearers of human rights—there is no non-coercive (and hence autonomy disrespecting) way to limit the heterogeneity we inevitably fragment into as we exercise that autonomy. Rawls recognizes this as a moral premise of modern life, what he calls the "fact of reasonable pluralism" that is the natural "outcome of the free exercise of human reason under conditions of liberty."[52] Given our situatedness in cultures and traditions that predate us as individuals, a subsequent tension inherent in the actual exercise of our autonomy is that we thereby also inevitably define ourselves through categories that we do not ourselves invent or, in most cases, even freely choose for ourselves. This is what theorists of intersectionality have long emphasized: if I am an African-American bisexual woman, I may choose positively to embrace each of these identities (i.e., "African-American," "bisexual," "woman"). I may even consider myself to be "proud" of them. But at the same time those embraced identities are largely historically constructed and presented to me as socially pre-fashioned. So even though I have indeed made these choices, I cannot be *fully* autonomous with regard to them because the menu of identity options from which I've chosen has been pre-fabricated and, in that sense, foisted upon me rather than being conceived and designed via my own unfettered exercise of will. If they exist this side of sanity, identities that are *wholly* chosen would be exceedingly rare (an identification with one's own literary creation?). Still, there would appear to be a spectrum of identity autonomy, ranging from the more foisted upon, what Sartre called "facticity," to those that are experienced as more deliberately chosen. An analog might be the difference between choosing among a subdivision builder's three pre-set floor plans and designing one's own house. Both choices involve some facticity and some choice, but designing the house oneself clearly involves *more* choice. In an ultimate sense all choice may be illusory (as in Spinoza's conviction that free will is merely ignorance regarding the causes of our actions), but there is certainly a greater *experience* of choice in some situations over others.

Originating in legal scholarship's critical race theory, the notion of intersectionality tends to focus mainly on "ascriptive" identities, which are those externally imposed upon us, including under oppressive circumstances.[53] In a sense the notion of ascription splits the difference between free will and determinism because it realizes that although most identities are in a sense fabricated they may be projected onto others with or without their assent. A good friend once told me that when he was a child he told his Mom that he didn't "feel Jewish" and so he didn't think he was. Her response was that maybe so but the rest of the world will not let you forget it; others will cause it to catch up with you. This is even more unrelentingly the case with identities such as "African-American" that tend to be more publicly identifiable. No matter how you "feel" about yourself psychologically, owing to your inescapably social and historical contexts, the world around you will commonly insist on seeing you according to *its* own terms, terms that are often artifacts of oppressive power relations. One can of course push back against the pressure to conform (existentialists like Sartre locate our freedom in this reaction), but it is not possible to live fully in this pushback mode; each of us represents some measure of acceptance and rebellion regarding ascriptive identities. The social machinery of ascription is of course often powerful, as in the labeling process that accompanies identity-generating situations such as disability, mental illness, a criminal record and many others. Once a person has an official record of X, Y or Z, there are circumstances in which others will make their own definitions of that person's identity, regardless of whether the person in question *chooses* to identify with X, Y or Z. Examples of this kind are legion and there are as many of them as there are oppressive circumstances that trap and pigeonhole individuals by socially-assigned ascriptive identities, "pinned by a look, like a butterfly fixed to a cork," as Sartre's Genet says about his childhood internalization of the identity "thief."[54] This reality is an important reminder that the putative liquidity of modernity does not imply a fantasy libertarianism of absolute freedom of choice without regard to the ways in which identity delineations are materially or societally *imposed*. Social realities can be monumentally inconvenient.

Yet as much as we are hemmed in by facticity and ascription and as much as the world pre-lays its own template for our everyday choices (e.g., consumerism), in liquid modernity we are also made to feel that it is imperative for us to author our own identities, to realize at least some degree of autonomy.

This makes us more "authentic" in the Kantian sense that in our choices we have not allowed ourselves to be played by anyone else as a mere means. Kant emphasizes the inherent dignity inhering in our autonomy and implies that it is independently valuable, this sense of authoring one's own existence, "I did it my way" sensibility. In retrospect this seems quite an appropriate stance for an eighteenth century that was transitioning politically from "subjects" to "citizens," or, parallel to that emergence, a slave who is freed where, like slaking a severe thirst, there is pleasure in the immediate reversal of his or her condition. But like many Enlightenment thinkers, Kant seems too caught up in the fanfare of autonomy to perceive the long-term acid corrosive power of that very same will-exercising autonomous subject.

With the benefit of a few hundred years of hindsight, a perceptive contemporary observer like Bauman recognizes that this delightfully free-floating subject whom Kant envisions is now subject to ailments characteristic of precisely such self-reliant free-floaters: ambivalence, anxiety, regret, depression, loneliness and so on. Each of these is symptomatic of the severe existential weight borne by individuals continually having to *choose* themselves for themselves, like a great leap that is initially exhilarating but then as it continues becomes vertiginous and frightening. Part of this is owing to the existential structure of choice itself, where there is a certain anguish inherent in any form of choosing, if only for the fact that by definition any act of choice is also simultaneously a destruction of alternative possibilities, paths not taken and opportunity costs. This is a paradox that is always felt at some level: the actual exercise of choice always obliterates choices; choosing removes options. In a kind of trick of temporality played by memory, the roads not taken can continue to beckon.

But even as choice is destroyed by choice, a new plateau of choice opens. This is how we avoid the existential hangover of choices: by not torturing ourselves with the inevitable regrets but "moving on" within the new menu of choices and making our next selection, *ad infinitum*. These are very powerful psychic forces. For Hannah Arendt something analogous explains the original psychological appeal of Christianity: the ability to be forgiven the inevitably but tragically uncontrollable consequences of the choices one has made.[55] To continue with the liquid modernity metaphor, one way to stay afloat on liquid is to maintain a certain speed as in, say, waterskiing; as long as you keep moving and maintain control, fine, but if you lose speed or stop you sink. This is one way to understand how placelessness can lead to identity proliferation. In the wake of neoliberalism, as the responsibility for everything devolves upon the individual— and the individual alone—there is a need to refashion oneself, to reshape, redefine, recreate; there must be a perpetual introspective motion, where one continues to surprise oneself and others with, say, the learning of new skills, the "discovery" of new interests, making new affiliations (e.g., the book group, etc.) or, in more austere imposed circumstances, getting fired again and landing yet another new job, adjusting to a new boss, finding oneself suddenly saddled with an unanticipated caregiving obligation (e.g., parent, child, spouse). Each of these alterations carries with it some degree of accompanying identity adjustment: that new reading interest makes one a "fantasy geek," the parent with Alzheimer's makes one an "Alzheimer's caregiver" and part of a support group, and so on. However much they might be imposed or chosen, the identity formations proceed with their own momentum.

Intersectionality's insight that we contain multitudes applies here. But the situation is significantly more complicated than merely enumerating my imposed and sometimes oppressive identities and neither is it just a matter of aggregating the consumer-like choices of "what I want to be." What is definitive of liquid modernity is not that we have a few interests and obligations or that as a matter of descriptive anthropology people play multiple kinship and public roles. It is that: (1) these identities are more and more experienced as *chosen* (even though the older kinds that are foisted upon us are still very much present), and (2) our adoption of these identities is more and more *individualized*; "it takes a village" is a nice saying but we are less and less involved in anything like a traditional village mentality across all spheres of our lives. For example, even at one level one would say one did not exactly *choose* to be in the situation of caring for Mom when her Alzheimer's progressed, one probably *did* choose to take on that role of personally providing the care as opposed to Mom going to a nursing home. (Even in the US, Medicare for the elderly makes this in principle an option for everyone if their parent's illness is officially recognized.) One may say rhetorically that it wasn't a choice because the nursing home option was always "unthinkable" but nonetheless it was of course actually a choice. And once the choice is made, one typically has to figure it all out on one's own. There are "resources" and so on available to help but the responsibility is on the individual caregivers (who most assuredly know who they are and feel the weight of the obligation) and not, at this point, on any "village." In the best case one gets enough help from friends and other family to make one's private psychological burden tolerable. But, especially in the US given the current state of the healthcare system, that burden seems almost inevitably privatized; caregivers often feel immense personal *guilt* about what they're able to do or not do for their loved one.[56]

That caregivers—often self-sacrificing and quietly heroic— should nonetheless feel such emotions as guilt is emblematic of liquid modernity's broad tendency toward self-determination, that is, enacting Kantian "maturity" by making one's own choices, while at the same time those very same choices seem determined by external circumstances not of one's choosing. I didn't choose for my Mom to have Parkinson's. (And neither did she.) So that caregiver identity is both chosen and not chosen. As Bauman expresses it, the "yawning gap between the right of self-determination and the capacity to control the social settings which render such self-assertions feasible or unrealistic seems to be the main contradiction of fluid modernity..."[57] We push our "self-assertions" yet they also pull back at us; we play them yet are also played by them. The foundational Kantian distinction between autonomy and heteronomy is severely challenged here. It seems both and neither.

This is what liquid modernity does with identity. There is the liberatory aspect of trying on something new or throwing oneself into a new and/or previously repressed interest. Yet there are the other kinds of identity choices that are not so exhilarating. But what is salient here is not the mere fact of any particular identity, however wrenching its presentation might be. (And it must be remembered that sometimes these are life and death, as many have committed suicide rather than face the fall from a particularly rooted identity, a human cost we see, for example, in the wake of every financial crash.) Rather, what is noteworthy is the sheer accumulation of all the identity choices, the very fecundity of the proliferation itself (again: parallel to the explosion of consumer choice). *So many* identities are on offer and there seems in principle no end to them.

This seems so overwhelmingly the case that those dealing only with commonly recognized identities like race, class, gender, sexual orientation, religion, ethnicity, etc.—the "official" identities to which critical theorists of intersectionality arbitrarily restrict their mixing and matching—seem at this point to be positively quaint. For people are living their lives by adopting an unimaginably diverse range of often intensely-held (in the sense that destabilizing them could be a life-or-death matter) identities that give meaning and coherence to their otherwise chaotic and/or empty-feeling lives. "Science fiction blogger" and "library volunteer" and "sexual submissive" and "parent of a schizophrenic child" and "holiday Santa" and "Cubs fan" and "body builder" and "volunteer fireman" *ad infinitum* may not seem as august and world-historic as the 1970s US Office of Management and Budget "racial and ethnic categories," but try and tell people that those kinds of identities are not important for their quality of life. One person might find another's identity silly (Otherkin[58]) or superficial (Katy Perry Fan Club) or weird (Juggalo) or dangerous (Psychonaut), but that's just it: it doesn't matter what *you* think. Like pain which by definition is "real" if it is felt, these identities *matter* to individuals simply because they matter to them. And it doesn't much matter anymore if their family or neighborhood or "village" finds them worthwhile or not; if one needs to one can, online or face-to-face (or both) depart at least *in mente*, or even face-to-face (as in the interesting case of the Juggalos),[59] from one's ascriptive community in order to find an alternative intentional community of the like-minded and/or relevantly congenial. Not only can one increasingly choose new personal identities but one can choose new villages too.

Social media has made it easy for those with decidedly minority interests of every kind to locate one another and then collectively to concentrate and intensify those enthusiasms such that they become more than just proclivities but "lifestyles" and identities. Emerging sexual identities provide good examples of this process. Various fetishes or kinks have existed forever, but were often experienced as isolated "perversions," often productive of searing personal guilt under the gaze of religious authority and custom. But now just about every sexual fetish one can imagine has a Tumblr page (or many of them) and countless devoted forums and other websites. Surprisingly lengthy master lists of emerging gender identities and sexual orientations are constantly being compiled online and it is easy for outsiders to gawk at the exotica on offer—and it is also impossible to know if some of these identities and orientations are truly internalized by actual individuals or merely conceptual possibilities. One

Tumblr compilation lists over 110 gender identities, including, just from among those starting with "a": "affectugender" ("those whose gender is influenced by mood swings"), "ambonec" ("identifying with both man and woman, yet neither at the same time"), "anogender" ("a gender that fades in and out but always comes back to the same feeling"), "autigender" ("a gender that can only be understood in the context of being autistic"), etc.[60] There are similar evolving sexual orientations including uncommon designations such as "Gray-A" ("do not normally experience sexual attraction, but can sometimes"), "objectumsexual" ("someone who is sexually attracted to inanimate objects"), and "skoliosexual" ("Someone who is attracted to genderqueer or non-binary people").[61] These lists are not necessarily accurate and the "reality" of individual items may certainly be contestable. It's impossible to know for certain. There are plenty of currently widely-accepted designations that would have provoked incredulity a generation or two ago such as "genderqueer" or "hetero-/homoflexible." The story here is not the *particular* gender identities and sexual orientations on the various lists but rather their very dynamism and quantity. That is the "liquidity" of liquid modernity in action.

Further flux is found not only with gender identity and sexual orientation but among what one might term "erotic identities" as well. For example, one popular social media-style kink site, fetlife.com, boasts an "options" menu containing over 80 possible defining "roles" (e.g., dominant, submissive, pet, little, primal, master, slave, kajira, kinkster, rigger, furry, exhibitionist, voyeur, top, bottom, Princess, masochist, evolving, etc.).[62] It is important to understand that, while they can be, these are not just sexual fetishes or occasional proclivities. Many people using these descriptors see them as much deeper expressions of their core identities, as containing crucial information about *who* they are vis-à-vis others. This is, to say the least, not our grandparents' dating scene. Aficionados describe themselves as into X, Y or Z fetish or kink and describe kink-less non-aficionados as "vanillas." Just as in the case of ethnic and other established identities, there is predictably even an internal politics of authenticity, where certain participants assert themselves as "lifestyle" X, Y or Z in order to distinguish themselves from mere touristic, thrill-seeking or "bedroom only" adherents of X, Y or Z. This is wholly analogous to the power move commonly perpetrated by religious sects or within regular identity politics where it is a strategy to achieve preeminence to claim authenticity qua "holier than thou" or depending on the group, more X, Y or Z than thou. The group dynamics are such that fanatical purity tends to be respected and made way for, as through the acquiescence of group members, certain individuals and cliques appoint themselves guardians of the lifestyle or identity. These definitions can occur rapidly. When the BDSM-themed *Fifty Shades of Grey* became a runaway bestseller (150 million copies worldwide and counting—presently tied with *Peter Rabbit* on the all-time bestseller list), it was common to hear contempt for all the "newbies" from those claiming to be more authentic BDSM enthusiasts, and self-appointed experts chimed in to lambaste the book and film.[63] These discussions seem to contain just as much intensity and display much the same group dynamics as the identity-generating theological or internecine leftist political ideology disputes of years past. And why shouldn't they? They are just as important to those individuals' lives. Again: it doesn't matter what *you* take to be significant or not, if certain individuals take X, Y and Z to be significant to their own sense of who they are, then X, Y and Z are *ipso facto* significant. Autonomous agents do not by definition need anyone's permission to interpret themselves as they do.

This is where modernity is really liquefying in a way that is undeniable as, despite the persistence of group-based oppression, core traditional identities such as religious affiliation have never seemed less stable while the flux occasioned by emerging identities has never seemed more diverse or less predictable. Who knows what oddities (from our current perspective) are emerging at this very moment? I guarantee they are out there.

My argument is that there are material conditions for this proliferation of identities in the wake of what Jean-Francois Lyotard (1979) famously called "the postmodern condition," which he elegantly and efficiently defined as "an incredulity toward metanarratives."[64] On my view these processes of incredulity are occasioned by a deterioration in the sense of *place* experienced by individuals as traditional familial ensconcements, where one reproduces one's elders' social place, give way more decisively under neoliberalism eventually to: (1) a period of lifelong at-will employment (i.e., where the employer contractually can terminate one's job in a way that, say, a feudal peasant cannot be "fired" by his lord) that cannot really be passed down inter-generationally but nonetheless often felt almost that stable, particularly under conditions of relative prosperity such as postwar America; and (2) an even more radical reshaping of human beings to the production process where they become wholly *discardable* as a function of increased efficiencies in the production process (e.g., automation and the like). I suggest that there exists an important link between individuals' economic placelessness and the proliferation of their identity options (and those options' attractiveness); however one may critique them, previous eras provided more secure identity anchorages. Much of this was oppressive, of course, e.g., religious persecution, as those anchorages were actively *kept* in place. But, as they say, life was simpler then.

Depending on one's situation, being rendered placeless can be liberating. Yet it can also be frightening and a cause for personal despair. And disorientingly it can be both simultaneously. One might be freed formally from religious persecution, yet in order to actualize that freedom one might find oneself an ocean away and cut off from almost everything one had known before.

Similarly, the array of options for self-definition can intoxicate as one feels the rush of freedom as one moves away from traditional expectations and moralities. Yet the past—via history, tradition and personal biography—always seems capable of exerting at any time a rip current that can pull one out to sea at unexpected moments, even during fair weather. Personal reinventions should not be expected to be without cost. So for the sake of rational self-preservation these intimate metamorphoses need to be *worth* it. Yet how is one to weigh the costs and benefits? Even if deliberately embraced, identities seem either to take hold or not, case studies in the Humean conviction that passion precedes reason. It simply does not seem that identity choice is "choice" in the sense of menu options; what presents itself as an identity "option" appears as such because of my previous inclination toward it. Such is the case with sexual orientation, for example, while it is implausible to suggest that individuals choose what gender to be attracted to, they may choose to *identify* as one among those who have either chosen the identity or have been ascriptively identified by it. (The latter case would be one where one might choose to "out" oneself in, say, a context of expressing solidarity with persecuted group members where one had the option of remaining undetected.) Among the parade of identities before me, even in this digital age, whatever my reactions to them it is clear that only a small portion of them are going to be actual live options for me.

It is an enduring mystery why this is so, why a person is inclined toward some options and not others. Even so, there does seem to be something like an "available range" of identities— perhaps we could call it an "availability spectrum"—that are possible for me given my internal constitution and my external social context, this latter in the developmental sense and also what is actually available in my present social field. Whatever this calculus of possibility for an individual, what seems impossible, though, to reference Robert Musil's classic novel, is to be a "man without qualities" in the ideal existential sense of resisting all determinations in favor of styling oneself toward an expression of radical negation, a pure negative freedom.[65] While particular ascriptions and self-ascriptions may be made poorly and in bad faith (e.g., racism) and so are always contestable, we always present ourselves *to* ourselves in certain ways. Identities are demonstrably alterable along the life course of an individual. But even during the pendency of their psychological succession there is no point at which *no* identity is present, no "dead space" of consciousness wholly devoid of self-determinations. Individuals surely experience vertiginous transition periods when realizing a wrenching "truth" about themselves, regarding, say, their sexuality or a religious "road to Damascus" or some other aspect of oneself taken to be at one's core. And perhaps this is as close as we are likely to come to an experience of pure negativity: the near-nauseating recoil as we struggle to regain our balance— if we are to survive, re-establishing *some way* in which we see ourselves once again. Doing without an identity narrative seems as impossible as doing without food or water; where there is human life, it is present.

And it strives always to reassert itself in the aftermath of change. The dynamics of that inevitable reassertion become all the more important in an environment where identities are in flux, where they become liquid. The new and expanding world of under-work and post-work is precisely one such environment. It is my contention that the shape of these reassertions provides the parameters for the shape of the human future. It is also a caution to all social planners of whatever ideological hue. The human need to construct and consolidate meaning—identity being merely our first-person window out onto that meaningful world as it has been co-constructed—will ultimately always trump rational self-interest, including that which might be identified as one's own best self-interest along economic or environmental lines. With these latter there is of course an intertwined threshold of survival (viz., food, water, shelter along with climatic and other habitability) that must be met on payment of extinction. But beyond that, history has amply demonstrated that human beings are eminently capable and sometimes desirous of undergoing tremendous suffering for the sake of their identities and their supporting systems of meaning. Martyrdom was not invented by jihadis; examples are recounted in every world religion and literary tradition.[66] One thing they all share is that the martyrs understand themselves as inhabiting some system of sense-making and adhering to some narrative that describes how their actions are understandable within a larger frame, usually a conception of life's ultimate purpose.

Rational social planners of whatever ideology—from neoclassical to neoliberal to socialist and communist—by design tend to overlook this "irrational" transcendent element inherent in the human psyche: our capability and even propensity to commit actions against narrow material self-interest on account of exigencies of identity and meaning. This is why those in a modern frame of mind are wont continually to be asking such questions as: "What's the matter with Kansas?" "Why do blue collar middle Americans vote for elitist economic policies?" "Why do the Islamists undermine their own societal development?" "Why does middle England vote for Brexit despite the (allegedly) dire economic consequences?" Or, for that matter, at the extreme end, why has anyone *ever*—from Thermopylae to Socrates to Thomas More to Thích Quang Duc[67] to Mohamed Atta (of 9/11 infamy) along with countless others cross-culturally and throughout history—chosen deliberately to end their lives for a cause? At the absolute opposite end of the spectrum from rational choice theory and calculations of material self-interest lies this meaning and identity-based willingness to die for a purpose that is external to the mortal self, a conviction poignantly captured by Martin Luther King, Jr's celebrated yet haunting maxim that "If a man hasn't found something he will die for, he isn't fit to live."[68] Now of course logically this does not entail that everyone dying for a cause is therefore "fit to live" (like Atta, many have martyred themselves for evil causes) but accepting a sentiment like King's does however entail no less than what it says: in the terms of my argument, one's life is not worthwhile if it does not involve an identity that anchors itself to something larger and more durable than one's mortal existence. It is a kind of paradox of mortality in a way: one's internal experience contains durable meaning only by reference to what is external to it.[69]

A pertinent ramification of this phenomenon is that identity matters in ways that are difficult for empirical analysis to capture, as those acting on the basis of a larger-than-self ideology are likely to show as behaving randomly or erratically or otherwise defying predictability from the point of view of a coherent set of "rational" assumptions about human behavior. Admittedly there are patterns. Knowing the premises of a particular belief system from which an actor operates, say, an Islamist teen who has been radicalized by a particular cleric, one might well be in a better position to make behavioral extrapolations. But this capability would not hold in general across the great range of human diversity—until such time perhaps as motivating systems of belief have been coded and mapped and made to develop along with the range of variables encompassing billions of individual human lives. Until that time, a time which may never come, public identity-motivated actions and responses will show up under vague black box headings like "political volatility," "resentment," "nationalism, "extremism," etc. and indeed "identity politics." And *ex hypothesi* as modernity becomes more fluid (i.e. more individuals become unmoored from their traditional ideational attachments), these hard-to-analyze identity gambits stand to become not just the spice added to the economic main course of world politics but the main course itself. We are already seeing this power along a range of phenomena from Brexit and European disintegration to

Trumpean anti-establishmentism to ISIS and rampant Islamism.

Liquid modernity ups the ideological ante. Because it is a natural feature of human consciousness to self-narrate, when traditional identity templates erode, ongoing mental repairs are instituted in both intimate and collective settings. After a nihilistic interregnum in which whatever idols experience their twilight, there is bound to be a reconstitution process that functions toward the installation of successor narratives. The stark ideological brutalism of modernity is as intolerable as staring too long at the sun; it gives no quarter to our need to place ourselves somewhere—anywhere—in the cosmos. In this respect, modernity's neoclassical core is built around a conception of the self that may be intellectually palatable as a self-interest maximizer but is at the same time impossible to embrace as a *lived* phenomenon that provides sense to all of the choosing behavior. We gain a great deal of predictive capability with this simplified model of the self but at the same time we lose an ability to narrate answers for ourselves to second-order *why* questions that are temporarily repressible but not durably so. At some point, because it is inherently insatiable and hence unstable, neoliberal consumer activity begins to introspect against itself and to detect a temporal pattern to all the choices. In this way a directionality emerges that brings to the fore questions about *who* is this self that is making these choices and *where* is it headed—perhaps on a road to nowhere. These questions are *literally* unsettling.

Chapter 2

Liquid paideia

And since the basis of education is a general consciousness of the values which govern human life, its history is affected by changes in the values current within the community. When these values are stable, education is firmly based; when they are displaced or destroyed, the educational process is weakened until it becomes inoperative.

Werner Jaeger, Paideia (1945)[1]

Exclusive inclusivity

I proceed on the premise that increasingly many people have been sidelined from productive economic activity, at least of the stable and formal kind that had characterized the retrospectively golden era of postwar capitalist growth.[2] This sidelining comes in many forms: from chronic unemployment ("discouraged workers") to drug addiction (especially opioids) and often comorbid mental illness, disproportionately race- and class-based mass incarceration, and populations who have been for whatever reason discarded as unfit for productive work. Isolating these sidelining processes can be elusive, but the overall trend, according to the US Bureau of Labor Statistics, is that the labor force participation rate seems to be on a downward trend, especially for youth.[3] In several of the MEDCs, youth un- and under-employment remains shockingly high (e.g., Greece, Spain, South Africa and many others), well over 50%. Even in wealthier countries characterized by a more laissez-faire approach and austerity public services cutbacks, we are increasingly experiencing an updated version of a perennial social problem that has always been regarded as a particular danger: legions of idle young men. Those in today's version of this neo-lumpen group are supported largely by others in their households and spend much of their time playing video games and in other "leisure" activities.[4]

Even for those involved in the workforce, their participation takes on a new character. No longer as in previous generations does one learn a vocation or skill with the expectation of landing a durably stable economic place. Rather, one is exhorted to become a "lifelong learner," now an official cliché of the education establishment that is sold as the emblem of some kind of liberated, spiritually swashbuckling life, while in reality this marketing gimmick merely designates a supine malleability toward perpetually retraining for temporary jobs. One now "wins" economically by rendering oneself maximally serviceable and fluid vis-à-vis employers' ever-shifting needs, by becoming in effect a bespoke human being. Diving in and out of as many work contexts as possible is presented as the truly fulfilling life, where one's ultimate allegiance is some sort of mix of self-regarding ideals with a libertarian self-help cast to it: delightful contemporary Candides for hire; individuals are personal "brands" in the service of everyone and no one.

One might think this would represent a boom time for educators, given the ubiquity of these self-development imperatives for "the Brand called You," where "you can't move up if you don't stand out."[5] Everyone needs in a sense to become their own little credential mill generator and their own little PR firm, in order to be able to engage in some ongoing and at least mildly tech savvy auto-didacticism. There is in other words plenty of "education" still needed for striving individuals who want to position themselves to get ahead, either through the acquisition of actual skills or through acquiring the credentialed social capital accruing to those possessing simulacra of those skills. In fact, one could argue that these are more important than ever for individual life chances. Higher education loudly trumpets statistics showing how much more those with college degrees earn over their lifetimes than those sad and formless un-credentialed ones consigned to the outer economic darkness. Learn and get paid. Or: *pay* to learn and get paid; that one should view college as an *investment* in oneself is now common sense.

But beneath this frenetic hive of economic preparation that allegedly characterizes our educational institutions lies an emptiness so vast that it is often not even noticed as such, similar to how those walking through a giant meteor impact crater might not be able to perceive their encompassing topography for what it is. Analogous to Pascal's "God-shaped hole" in our modernized hearts, there is at the core of our institutions of mass education a growing void.[6] The justifications, rationales and *raisons d'être* of yesteryear have expanded and become laden with complexity. But this rise in complexity is like a baking crust that has risen to the point of becoming thin and brittle, containing hollow cavities beneath into which it could collapse under the slightest pressure. For by far the main ingredient in the public justification of today's schools is the simple expedient of economic utility, both for individual enhancement of career chances and also for society as a whole in terms of enhancing economic dynamism and global competitiveness, the latter being assumed to redound in the long run to the collective benefit of the taxpayer-investors who fund the scheme along with the student-customers. Education is therefore an investment that like any other must yield dividends. It must *pay*.

There remains a ghost whisper of broader civic ideals a la Jeffersonian democracy, but this tends to be so muted as to be nonexistent. In certain religious schools, too, there remain vestigial ideals of a more comprehensive nature that gesture toward something beyond mere economic gain. But on the whole the schooling machine runs on the justificatory fuel of its promised economic payoff. It has probably been a couple of generations since a major US politician—from left to right—last mentioned education in any terms other than those of economic utility. Aside from regular but empty ritualistic nods at civic education (and even these are decreasing), the only robust examples that come to mind are depressingly idiotic criticisms by Christian fundamentalists that the abandonment of school prayer and the teaching of evolution have resulted in schools' steady moral decline. Particularly in higher education, the economic rationale for inclusive education has morphed away from a more collectivist notion of economic nationalism, where "we" will all be wealthier by educating our society as widely as possible. The idea here is that all those future inventors and entrepreneurs and so on — what we turn up when we rake annually through the Jeffersonian "rubbish" — will in turn strengthen the collective due to the economic value they will add.[7] In the manner of the invisible hand, though, this collectivist economistic rationale has given way decisively to a more individualized consumerist economistic rationale where the point of school is to better *yourself*, protect *your own* future earnings and social position, and so on. This individualization is yet another symptom of liquid modernity. One must above all invest in *oneself*. Investing in the larger society is still mentioned but is understood more as an aggregation of individual life projects rather than in any kind of first-person plural sense of a "we," that is, a collective project of the nation or the (human) "race" or however one wants to put it. Education is a tool to get ahead, full stop, and this has become all but obvious common sense. This is what the hard-headed educational managers at universities and elsewhere feel that they "know" and it is what makes them "realists" as against those among their faculty who childishly persist in thinking that teaching and research have value on their own terms and not merely as marginal cost calculations and public relations "buzz."

A couple of points to emphasize: (1) Economic utility is not by itself a substantive ideal, unless one possesses the traditionally-denigrated miser's worldview wherein the hoarding of wealth is held to be an intrinsic good (King Midas gets his comeuppance and Silas Marner is reformed[8]); typically even the most one-track economistic minds will allow that wealth is merely an *instrumental* good that is to be valued as a function of what it can secure, and (2) in most MEDCs, norms of nondiscrimination against cultural and religious groups require government schools to be largely neutral—and indeed often scrupulously so—regarding religious and other comprehensive worldviews, along the lines of the Establishment and Free Exercise Clauses of the US Constitution or slightly looser analogs such as those found in the European Convention on Human Rights.

There are nuanced differences across the MEDCs, but in larger perspective—compared to what might obtain in, say, a theocracy—the broad similarities are more telling. Two specific examples among many of this pronounced trend toward symbolic neutrality are presented by two court cases from areas of holdover religiosity within MEDCs, the southern US and Italy. Active religious practices have long been abandoned in public schools, especially since the 1960s in the US.[9] Going even further, both the US Supreme Court and the European Court of Human Rights have held that religious icons may not even be *displayed* in public school classrooms, neither as, allegedly, "moral education" (the Ten Commandments in Kentucky) or "cultural heritage" (crucifixes in Italy).[10] While these matters are not hugely significant by themselves—despite the local outcry in response to them—there are many analogous legal judgments. In their ensemble, these legal decisions symbolize a larger trend in which educational institutions across the MEDCs have become incrementally less tethered to any substantive animating ideals or anything like a Rawlsian comprehensive conception of the good. We have largely said goodbye to all that. The default policy problem for political liberalism *had* been how to moderate and adjudicate among fervent believers who were assumed to be thick on the ground. Though they still survive with strong voices in certain pockets, what we are now glimpsing is what it looks like when the presence of those believers—any types of believers—can no longer be taken for granted. Neoliberalism's wake thus leaves us not only with a spreading economic precarity but also with what seems a widening *belief* precarity as well, where traditional worldviews are dying without heirs. Those solids keep melting.

This grand trajectory makes contemporary MEDC education systems on balance *apaideutic;* they lack a paideia. *Paideia* is one of those notoriously untranslatable classical Greek terms that, following the educational historian Werner Jaeger, signals a conception of education as having to do with a pattern toward a cultural ideal, in Jaeger's words, "the creation of a definite ideal of human perfection." This constitutes, in other words, an animating ideal for educational efforts where everyone involved—teachers, students, parents, administrators, citizens, etc.—acts in concert under the aspect of a shared vision for what constitutes "the educated person" and how such a creature would connect to a vision of the good life as a whole *(eudaimonia)*.[11] Norms of inclusion and non-establishment, however, require institutional neutrality regarding comprehensive ideals (viz., religion), principles that are necessary for modern diverse societies. Yet they come at the cost of rendering public schools mostly incapable of inculcating shared ideals of life beyond perhaps the thinnest and most obvious of ethical norms (e.g., "wait your turn in line," "do no physical harm," "be on time"). These thin ethical norms are sometimes quite important, for example, norms of inclusion regarding race, gender, disability, etc. However, there is no guiding rationale available for exactly *why* these norms are desirable, no ability to integrate them into any larger narrative about the meaning of life and hence they become merely free-floating feelings, mere emotional impulses that may be turned into many different directions, including into an increasingly psychologistic "politically correct" rhetoric now afflicting college campuses. Youthful self-righteousness that once may have been directed by Mao or Hitler or religious zealots is now directed into denunciations of "incorrect" thoughts and statements and a pronounced impatience with free speech norms manifest in such phenomena as the maintenance of "safe spaces" and hypertrophic introspective attention to one's own feelings and "empowerment."[12] It is literally a farce: a cartoonization of moral conflict with everyone wearing simple masks to indicate their assigned role. This clumsy and emotive thrashing about is a symptom of an educational environment that has not for generations fostered any imperative toward the integration of "righteous" impulses with any larger world view; in such cases all that is left is the lone self and how it "feels" (about itself), a maddening self-referential loop that never really exteriorizes, that is, never alters anything outside of itself because its own internal feelings are the focus of its activism. They may be lions on their own campus and vis-à-vis the managerial nihilists maintaining the safe spaces, but to the larger society they are mere items of curiosity, amusing and/or annoying sideshows with very little appeal. Up close it can be difficult, but narcissism is one of the easiest things for people to spot at a distance.

The reason for this lack of crossover appeal—a liability shared by would-be "public intellectual" professors and firebrand students alike—is that the communal cultural ideals where one might find a paideia lie splintered in a thousand pieces. There is simply no larger shared story against which one's moral posturing may be seen as significant or, really, *as* much of anything at all. It's not even as if one's campus activist performance is seen as wrong or evil or in any sense truly provocative (one of the most curious features of the academic bestiary is the self-styled "provocateur" who never manages actually to provoke anyone). It's that it is not really *seen* at all, it doesn't register as much relevant to the life of anyone who is outside that immediate campus environment. There is no connective ideological tissue through which to make such an identification of interests. Within higher education, the cultural ideals presently on offer are merely shades of their former selves, hollow offerings like "global citizen," "conscientious consumer," "ethical/non-discriminatory person," "rights-bearer" and related incantations.

The bleak cornucopia of corporate lingo spilling from university web pages is ever more voluminous and self-parodying. At my own University of Delaware we endure public relations campaigns everyone is supposed to find inspiring, buffoonish yet vaguely menacing declarations that we are "thought leaders" who "dare to be first" while we devotedly practice "inclusive excellence."[13] Such marketing inanities are ubiquitous and many are even worse. Many are worse. The University of Buffalo (NY) re-launched its own marketing efforts—"a new identity and brand strategy"—with this cringe worthy bit of advertising copy under the banner heading "We Take the Lead." They elaborate: "The University at Buffalo is a vibrant and inclusive community of big thinkers and even bigger doers. We work together to question and upend theories, lifting each other up and driving change. Because at UB, ambition is a virtue, tenacity is a given, and discovery happens in the lab, in the lecture hall and everywhere else. *That's just how we do it here* [emphasis added]."[14] What a fever dream of roiling intellectual ferment is taking place in Buffalo! Who knew? And it is of course far, far different from what is going on down the road at Rochester or Syracuse.

This moronic marketing blitz is not unique to the above-named institutions and is multiply symptomatic of larger trends. First and most obviously is the individualism inherent in the personal investment model of higher education in which the student-consumer derives future earnings from the credential. The educational experience itself is also relevant but at least at the undergraduate level that part of it seems doubtful or at the least oversold.[15] It's main function appears to be what economist Bryan Caplan terms an external "educational signaling" (to employers) of the credential rather than any internally-acquired knowledge or abilities. These signals are what ultimately place graduates into their duly credentialed labor pools. On the job market, then, as Caplan suggests, "[c]onventional education mostly helps students by raising their status."[16] Thus arises the figure of today's increasingly desperate white-collar *chancer*, the get-ahead college student with suitably (over) inflated resume, standardized test-taking skills and fail-safe interview techniques. None of these self-marketing "skills" have much to do with the actual on the job activities but as requirements they have everything to do with the all-important placing of oneself in position via the educational signals.

This phenomenon of educational positionality is the not-so-secret hidden secret of the artificial market for the higher education credential and it explains much of the higher education marketing hype. The phrase "positional good" was first advanced by economist Fred Hirsch (1977) to apply to situations where the value of a given good is a function of its social ranking, i.e., the *perception* of who possesses it and who does not.[17] So for example the economic value of a college degree would be inversely proportional to its distribution among the relevant population; if everyone has that degree then its relative value will decline. Education has often been considered a quintessential positional good that is more valuable the extent to that it is relatively scarce. (If everyone has doctorates then all of a sudden PhDs are driving Ubers.) In a democracy with norms of universal education, this dilemma traps everyone into a credentials race which theoretically benefits everyone, as society's overall intelligence quotient is supposed to rise. Socially optimistic educational theorists of the previous century like John Dewey often seemed to speak as if the overall amount of "social intelligence" in society would be salutary in this fashion and would automatically yield progress that would be evident in more rational and smarter social policies.[18] This universalizing optimism is implicit in a brochure-ready mainstay assumption of higher education marketing campaigns: the more universal the college participation the better. However, one does not need to be a labor economist to appreciate the positional point that the ubiquity of the college degree debases its currency; the market glut causes the degree to degenerate from being: 1) a job guarantee to 2) a requirement to 3) an assumed precondition like basic literacy.

Economists Robert Frank and Philip Cook have pointed out that it also creates a "winner-take-all" dynamic that further refines and differentiates the signaled credential.[19] Whereas previously a college degree by itself might have landed one that desirable job, now that degree must be from a good or maybe elite school. The question becomes not "do you have a college degree yes or no?" but rather "*where* did you get your degree?" The winner-takes-all dynamic also causes smaller and smaller differences to yield larger and larger benefits, e.g., relatively speaking the difference in admissions cut scores between Harvard and UMass might be quite small but the gap in graduates' future earnings is nonetheless disproportionately huge. Under such intensified conditions, what matters most about education as a positional good, then, is that it ensures graduates' all-or-nothing *proximity* to revenue streams, and their consequent ability to siphon out of the money trough; frenzied or not, it is jockeying for position to *feed*. Matters only become unmanageable when so many incoming job aspirants crowd around the trough that the siphoning process is disrupted or, worse, if the expectant credentialed invitees cannot get close enough to "dip their beaks" at all. This is the point where the winner-takes-all dynamic must be readjusted and heightened, to thin the herd and reallocate the number of available troughside positions. This is another way of increasing productivity, and other efficiencies put increasing pressure on the education-jobs nexus. Ask the indebted art history graduate barrista to explain how this process feels as she brews your latte.

So the educational individualism has both temporal and spatial dimensions: as an investment over time in oneself and as a positional good vis-à-vis others. As such it is experienced almost as an immovable object by those within the credentialing networks. Educational time can be wasted and opportunities missed. Where is the bottom-line payoff of that humanities course? The opportunity cost of not enrolling in that more marketable major? All the while, officialdom is growing impatient with anything other than a nose-to-the-grindstone vocationalism. *The Chronicle of Higher Education* reports a notable instance of this impatience involving the Governor of Florida:

> Public higher education is dying. As senior faculty retire, their positions and programs are going with them, not to be replaced. There is always talk about closing "expensive" departments: the humanities, the arts and the social sciences in particular. Gov. Rick Scott of Florida recently declared that anthropologists were not needed in his state: "It's a great degree if people want to get it, but we don't need them here." On another occasion he said, "Do you want to use your tax money to educate more people who can't get jobs in anthropology? I don't."[20]

Bashing allegedly useless college majors has become something of a bipartisan sport in US politics, from the Republican Governor of Kentucky's weird objection to French literature to Democratic President Barack Obama's dismissal of art history.[21] All of this is a natural consequence of the student-as-consumer becoming much more protensive and vocational, in short much more receptive to seeing herself as a consumer and commodity in an education market. It is all about *her* future, *her* potential, *her* singularity, *her* greatness-to-be. In this frame, there is no overarching ideal outside the ego-borders of that self, no collective project with which one can identify regardless of one's role, no "bad" trough positions from which an individual might be discouraged; there are no evaluative criteria for moral judgments here and it is enough to be alongside the trough—*any* trough with a revenue stream. Success, as they say, becomes its own reward.

Into this setting enters the narcissistic moralizing that serves as the ethos of the preening and precious customer-student "snowflake," to borrow the widespread term of opprobrium from the political right. Here we have the left-wing of neoliberalism posing as an ostensibly egalitarian and inclusive identity politics. Identity politics is a natural stance for these student-consumer commodity hybrids, as it implicitly recognizes—rightly— that meritocracy is largely a sham (save for a few technicians perhaps) and that what is really at stake in the higher education credential factory is socio-economic *position*. This is what left-wing "political correctness" on college campuses gets right: despite how educational institutions sell themselves, economic reward is less tied to particular educational value-added skill sets than it is to opportunities regarding who gets slotted in which spot. When one is positioned near certain revenue streams, unless things go anomalously wrong, one pretty much has it made. The now-archaic discourse of merit and competence is decried as racist and sexist but its real crime is that it is irrelevant. To be sure, there are a few authentic talents—and these are showcased almost desperately in higher education marketing—but incompetence and do-nothingism among elites is too widespread and embarrassingly obvious in the Age of Trump while, conversely, talent and drive is too manifestly wasted and unrewarded among the precarious poor.

These phenomena are easily apparent to anyone with eyes to see them. There are many housemaids and pool cleaners who are far more intellectually curious and driven than wealthy college students. Of course the glaring lack of concern with economic class among the identity crowd has led only to a "diversity bureaucracy" which is, as bureaucrats typically are, far more concerned with self-perpetuation rather than the achievement of any external aims—whatever the official rhetoric.[22] They may be deluded as individuals, but their collective strategy is rational: simply *get in position* above all else, in part by ensuring that previous position holders "check their privilege," i.e., relinquish their reserved trough positions. If this is the game, any prejudices or exclusions that could impair future positionality are correctly identified as *the* obstacles to social progress. Insofar as positional processes are how societal resources are gate kept, the trough jostling is indeed a basic issue of civil rights and substantive due process, especially when it involves historically discriminated-against and federally-recognized "suspect classes" like race, gender, ethnicity, religion and so on.[23] But it only goes so far.

The focus on positionality implies a process-oriented view of social injustice. If we fixed the obstacles to institutional access, who it is that gets accepted to college and into the best majors and that sort of thing, then we will fix the injustices. Such a conception implies that the basic economic structure is acceptable and that the problem of injustice and inequality lies in the systemic maldistribution of position (and hence reward) within it. Economic processes are naturalized and the ensuing hierarchies are only unjust if they insufficiently mirror population percentages across whatever are designated as the favored identities, which in left political liturgy usually means the traditional intersectional concerns of race, gender, ethnicity, sexuality, etc. It is as if the key to political legitimacy were to lie in the diversification of hierarchies that are themselves assumed to be quantitatively stable. As Mark Dudzic and Adolph Reed explain, "From that perspective, the society would be just if one per cent of the population controlled ninety-five percent of the resources so long as significant identity groups were represented proportionately among the one per cent."[24] As long as the inhabitants of "The Capitol," as depicted in *The Hunger*

Games's desperately inegalitarian world of Panem, are culturally and gender diverse, then social justice has been sufficiently achieved.[25] This inclusive exclusivity is the implied dream of contemporary higher education managers, perhaps the one thing students, faculty and administration on most MEDC campuses tend to share in common. It is a kind of "my-turn-ism" that is, one must concede, a rational strategy for an outsider who wants to get in on a certain game. It is not always advisable simply to try to destroy the game and perhaps that won't work. Why not simply try to get a seat at the game table?

One additional benefit enjoyed by a more diversified elite, particularly within complex hierarchies of position such as large corporations and academic institutions, is that viewing them through diversity lenses makes the elite group appear more legitimate, perhaps even meritocratic. If elites appear internally heterogeneous, anyone bearing the mark of one of the "insiders" yet who remains on the "outside" has been left there due to her own shortcomings. Now some individuals do have shortcomings. But it is easy to see the ideological work being done here by the salaried diversity cadres: they are erasing economic class while signaling other virtues in order to redefine progress. All of this is accomplished while simultaneously reassuring everyone that their secure positions in the hierarchy are justified because, on the basis of the evidence of the diversity all around them, they must have acquired those positions as the result of an open and fair competition. This veneer of legitimation helps secure elite positionality and is thus an added attraction of elite diversification.

The effectiveness of this justificatory strategy is enhanced, as is the case currently in US academia, if the official narrative of the putatively diversified elite is policed to a sufficient extent by an army of diversity bureaucrats and other HR-style compliance enforcers. The identitarian *summum bonum* of inclusion (the telos of which is always left unaddressed), while of course laudable as

91

far as it goes, is quite obviously utilized as a diversion from that which actually threatens elite power, viz., class and meaning-based critiques (the latter involving the nihilistic elite's inability to maintain a symbolically sustainable lifeworld for the majority). For example, I have sat through numerous meetings with colleagues agonizing about the difficulties with "recruitment and retention" of students and faculty of color when the glaringly obvious solution of lowering tuition fees (or their equivalent) for them apparently *cannot* be mentioned. The reason? It's always beyond our faculty control in some fashion — whereas presumably the inclusion of everything else is somehow under our control and any shortcomings are presumably due to our personal racist wickedness. As Reed and others have suggested, this is how the managerial antiracism and identity politics function as the left-wing of neoliberalism, imbued with an individualist theology of sin as a personal character flaw and corporate "remedies" as salvation. In this respect, the HR-style mode of pursuing social justice is wholly complementary with that of the laissez-faire right-wing, as both join together in an ensemble of virtue-signaling with the ultimate effect of legitimizing the status quo where above all elite positionality is conceived as *deserved*. This approach is more a set of psychological techniques like cognitive behavioral therapy than it is a politics; it is moral self-help administered by professionals and performed largely for the self-interested purpose of institutional career advancement and institutional public relations, along with the abovementioned added benefit of a kind of narcissistic reassurance needed by the guilt-ridden winners in the positional sweepstakes. Student call-out protesters at elite universities (where these kinds of protests mostly take place) are unsurprisingly working in their own enlightened self-interest when they advocate for greater inclusion in their inherently exclusive arena.

More, given human narrativizing tendencies and our own peculiar cultural traditions having to do with sin and the like, there is a need for the well-positioned to justify their own ascension to themselves, to conceive of their life course as not only one in which they have done *well* but also that they have done *good*. Some individuals may at times actually see themselves as bad or evil (though this is debatable), but such a self-image is not a sustainable identity for inherently social creatures. And when it occurs it is often highly unstable and dangerous, terminating in self-harm and/or harm to others; truly seeing oneself as evil is in this way typically a precursor to literal erasure. So the tendency toward self-justification and rationalizing one's positionality can be a matter of moral survival; it can be hypocritical but we should be grateful for it too.

The implied long-term goal of this intra-elite inclusiveness panic seems indeed to be a Panem-style diversified yet otherwise *inclusively exclusive* elite, essentially a legitimation strategy for a new and improved ruling class. This probably does admittedly have the effect of taming that ruling class to an extent and sensitizing it regarding the genocidal tendencies of yesteryear and this is not a small matter. But the augmented meritocratic appearance also allows it to style itself as the sanctified product of a rigorous competition of cognitive and moral fitness. (One might speculate that this is a source of the special animus of educated elites for Trump with his pursuit of their best economic interests through massive tax cuts and the like: the economic and political ascension of someone so boorish indirectly gives the lie to their own meritocratic self-image.) Though annoying, it should also be admitted that this self-congratulation is less gruesome than cruder forms of eliminationism and so, in the context of the bloody twentieth century, it could be described as a kind of advance for humanity. Meanwhile, as the shrinkage of the middle class lends the question of economic positionality a starker all-or-nothing character, the urgency of legitimating the winners and the losers grows proportionately. Due to these material conditions it is no surprise to see an increasing hysteria among corporate and academic elites concerning their internal heterogeneity and inclusiveness. The more they have of these two things the more justified is their hegemony and, ultimately, their predations upon the vast majority of the population, the deplorable ones inhabiting flyover country — "the poors," who are slated for elimination. And all the while the halos of the well-positioned remain untarnished; not only do they get the wealth and the power, they also get to be "the good ones" too. Ideological heirs to Calvinistic anxieties, their place atop the social hierarchy is itself evidence of their salvific worthiness.

The political flaw in this self-understanding is that, unlike the original milieu of Calvinism, which was connected to a theology and a comprehensive conception of the good, this time nobody beyond the allegedly elect finds their mirage meritocracy convincing. Modern universities' technological research achievements can still sometimes dazzle and enrich certain segments of the population but, as everybody knows, the university personnel directly involved in those kinds of achievements is a tiny percentage of the population of the academic *Gesellschaft*. A casual observer is forgivably often mistaken on this point, as every university markets their scientific and technical research especially loudly—in the US alongside marquee intercollegiate athletics—to be tangible evidence of public benefit. We are smart, beautiful and morally exemplary technological wizards, excreting pro bono solutions to everyone's problems; "everything we do, we do for *you*." This is the distilled PR script.

Higher nihilism

In official pronouncements there is pabulum about *people* being the university's main product or resource but it is clear that the public is to be led to value State U. because of its *benefits* or, in the managerialist lingo *du jour*, "impacts," of the teaching and research, especially on the economy. ("Impact" as a criterion for university academic assessment is particularly big in England but is spreading everywhere.) This makes perfect sense in a corporate environment where impacts are clear and relatively unambiguous, namely, sales, profits, etc. A conscientious company could even add environmental impacts and other externalities to what it considers relevant. Impact also makes sense in areas where there is a high level of consensus (spoken or unspoken) about what outcomes are considered desirable, such as in medicine or public health. Yet as the managerialist drive toward uniformity inevitably sweeps across the heterogeneous research landscape of the multiversity into less immediately tangible areas like basic science and mathematics, the social sciences, arts and humanities and education, matters become foggier. There is a quick way to gage the emptiness of impact rhetoric in one's home department. I have done it and it is simple and enlightening. When discussing research criteria and how to incorporate "impact" into, say, one's promotion policies, raise a simple question about what *kind* of impact should be sought. For presumably what is desired is that the research have a *beneficial* rather than a deleterious social impact; the outcomes should be *good* rather than bad. Yet this simple question is difficult to answer as the level of societal consensus weakens regarding outcome. As stated above, while one can still have debates about what exactly constitutes "health," medical researchers and many in other applied sciences don't have to worry about this so much as they can assume that saving lives, curing diseases and reducing the medical and business costs of illnesses, etc. are obviously welcome.

In hybrid areas like education or criminal justice one sees how things begin to break down. Everyone wants better education and less crime. Yet although the research in these areas is funded largely on the medical model, "the data" upon which research in these fields tends to proceed is highly contestable. In K-12 education research, what stand in and pretend to be analogous to health outcomes in medicine are usually standardized test scores whereas in criminal justice this role is often played by official crime statistics. In a grand game of The Emperor's New Clothes, to keep the research grants rolling in, everyone pretends on this basis that their work is "data-driven" and hence respectably "scientific." The only problem is that large-scale social phenomena like education and crime are too complex and multivariate to allow much to be concluded from any one intervention.

Conscientious researchers in these fields will respond that their modeling can control for enough of the variables to allow for conclusions to be drawn. This is true. However, controlling for variables (say, geography, socio-economic status, race, class, gender, age, birth order, weather, language, etc.) *by definition* narrows the generalizability of the research, in principle to a vanishing point. In these hybrid *wannabe scientific* fields (i.e., those like education whose data are *themselves* complex social products rather than the simpler observables of hard science), the logic of conscientious research purchases precision at the cost of usability or, ironically, one might say, potential *impact*. Thus in terms of determining causality (which is the goal) one can almost never say anything very generalizable from any of this research; conclusions must always be so carefully and elaborately qualified that at best one gains "suggestions"—and so the scholarly enterprise here consists of suggestions piling up upon suggestions. As a result, these fields are structurally beset with fads like "new math" in education and "broken windows" policing in criminology that decades later become embarrassments. Practitioners in these fields are typically blind to these dynamics because success in their research has come to be defined in terms of annual government grant funding competitions and winning these competitions is what is rewarded not because of any appreciation of the actual impact of the research itself but because of the research dollars the grant awards contribute to the university's bottom line. Thus the only real and lasting *impact* consists in the cash procured for the university. In my career I have witnessed this process deepen. At faculty meetings we used to applaud the research itself. Now the most enthusiastic applause is for the announcement of research grant dollar amounts. Faculty actually sets aside time now to applaud money itself—and I mean *literally*, to cheer for it, the higher the amount, the louder the cheer. It is strange to see such displays within academia.

In this context, raising normative questions about impact makes one the skunk at the party. If as mandated we are to reward impactful research, should we distinguish between *good* impacts and *bad* ones? What if the researcher up for promotion has had loads of impact but the impact of her research has been *bad*? (For argument's sake, let's say we can tell what the impact has in fact been, a capability which is, after all, implied by the very insistence upon utilizing the "impact" desideratum). Are we to reward and otherwise promote research with bad impacts as well as good? If yes, what do we make of the entire enterprise and our PR claim to be helping society? If no, what are our criteria for making the normative distinction? The response I have personally encountered from such questions is silence and nervous laughter proceeded by an expeditious "moving along" to further agenda items. The laughter possibly may be due in part to my own off-putting personal qualities but I suspect that this is not its sole cause. Faculty meetings are dull but they are fairly congenial settings and we typically discuss many things and have famously digressed many times over the years. The silence, I think, is because everyone instantly realizes that the normative question about impact is, practically speaking, *unanswerable*. Consequently, to make any headway would require a much larger discussion than anyone wants at such moments, something resembling a *philosophical* discussion about what is in fact good and bad and what is worthwhile in life and society generally.

Philosophers may love moments like these but normal people just don't. And for good reason: in a public university in a heterogeneous society it is highly unlikely to win any consensus on deep moral issues and therefore that discussion is likely to lead to discord and dissensus. At best it would be a waste of time as no conclusions are reached and at worst it could lead to group dysfunction. In the particular institutional setting of a faculty meeting it is *sensible* to avoid such talk and to ignore the would-be gadfly.

But this politesse comes at the cost of an implicit collective acknowledgment of the vacuity of impact talk. The vacuity becomes especially apparent when scholars in humanities fields are asked to play the impact game too. In the social sciences there is a *prima facie* focus on "real world" social problems and the empirical research methodologies, especially in qualitative research, commonly involve observing and recording the doings of other flesh and blood human beings. In the humanities the "data" tends to be textual (e.g., archival material in history, *maîtres á penser* in literary criticism and philosophy), so it is one step further removed from those in one's immediate environment. In such contexts, any impacts—if they can ever be detected at all—are going to tend to be very long-term (perhaps even posthumous), less direct and less tangible. In a pluralist environment, by which I mean one in which there is not much that is shared in terms of ideological assumptions and/ or a comprehensive conception of the good, it will be almost impossible to detect any societal impact of such work. If it can be meaningfully measured at all, the impact of a humanities faculty may be detected best in its own teaching and, in rare cases, the readership of its articles and books. Even if, *per impossibile*, these kinds of impacts could be reliably ascertained, the normative question would still remain and in fact would even be more fraught: what about *good* impact as opposed to bad? More, such questions probably are in principle unanswerable without reference to some axiological framework, a shared background upon which a governing comprehensive conception of the good provides normative criteria.

But in a religiously and ideologically heterogeneous society this psychical cohesiveness is precisely what is lacking. Imagine a small and cohesive religious school in late medieval times, a monastery say. Oddly, it is far easier to imagine a coherent discussion about scholarly impact in such a pre-modern setting because one could presume a greater amount of background agreement about what constitutes a bona fide contribution to the shared vision of the order. It will be comparatively clear to their fellow monks how the scribal labors of Brother Gildas or Brother Rabanus contribute to the guiding vision. An exquisite illuminated manuscript or a biography of a Christian martyr would glorify God in a way that is to them obvious. ("Glorify God" would be an "impact" whose desirability nobody would question although no one could measure it.) If, however, unlike the monks, there is no recourse available to a shared conception of the good, "impact" can be nothing other than a hollow and morally neutral concept. At best it may be able to count things but even then it cannot say what things should *count*.

So without any background comprehensive vision it is impossible for a university to make even the most basic judgments about the benefit to society of a core function like research. What tends to happen is they are reduced to a "brag sheet" compiling numbers of graduates employed, their earnings, research dollars accumulated along with a few camera-ready corporate "partnerships" with community organizations. Add in a few very practical and readily understood science and technology research products, along with a few other high profile projects for color, and it's a PR wrap. By necessity the research benefits are discussed in neutral quantitative terms such as patients cured or regional economic benefits or national defense contributions, all the while straying as far from possible, corporate style, from anything potentially normative and hence potentially divisive and alienating to any of the stakeholders. Whatever the official rhetoric, a school's actual mission becomes at this point mere self-perpetuation and quantitative growth in the corporate manner. This corporate neutralism is the diametric opposite from *paideia*, where a specific set of cultural ideals serves as a regulative ideal and thereby also provides a ground for legitimating normative judgments, as would for example the theological vision of a monastery or the ideological outlook of a party cadre school.

Like the modern business corporation, the modern university has come to stand for nothing beyond *itself* and correspondingly, in their aspect as participants, the students-customers are placed in a position in which they also stand for nothing beyond *themselves*. Multi-level self-regard then becomes the reigning common sense. Students quite sensibly ask, what can this college credential do for *me* in my career? Very far from any Socratic pedagogical *eros* (the very mention of which might get one fired), that is, any passion for learning for its own sake or for the sake of anything larger than tangible personal benefit, the internal relationships among the component parts (e.g., professor-student) become wholly transactional. At best there obtains an equilibrium of economic mutualism, a multi-faceted balance of self-interest between professor-student, administration-faculty, etc., what political philosophers call a *modus vivendi*, a platform for rational cooperation among overlapping interests, in this instance material ones.

The covert presumption is that the institutional machinery can continue running indefinitely via this kind of internal momentum among the self-interest maximizing component parts. Everyone can act egoistically as individual self-interest maximizers who by design locate the *telos* of their self-interest outside the institution; the framing is that the school *serves* those interests whatever they may be: be the best X you can be, whatever X you happen to choose. Under such conditions it is no longer necessary for the institution to have any substantive mission, a state of affairs attested by the genericity of institutional "mission statements" populating every university home page.

By contrast, nominally religious institutions such as Catholic universities still have vestiges of their originating substantive worldviews and often this difference is apparent vis-à-vis the mandated neutrality of public universities. For example Neumann University (Aston, PA) states out front that Christ is "the Source of All truth and Truth itself" and community members should embrace the "challenge to live a life rooted in Christ." Within that broad tradition it further specifies its core goal is to "Demonstrate a firm commitment to the Catholic Franciscan tradition" and derives its motto "Veritas-Caritas" from Ephesians 4:15: "Rather let us profess the truth in love, and grow to the full maturity of Christ the Head."[26] Notre Dame University (South Bend, IN) is strikingly forthright as well, emphasizing that the University "pursues its objectives through the formation of an authentic human community graced by the Spirit of Christ" and maintaining "a special obligation and opportunity, specifically as a Catholic university, to pursue the religious dimensions of all human learning."[27] Often religious universities come across as a bit more generic, for example Sacred Heart University (Fairfield, CT), references "the Catholic tradition" but then proceeds to list more or less boilerplate ideals that would be at home with the most diluted public university mission declarations. Sacred Heart's "vision" is that it "aspires to achieve prominence through innovative teaching and learning while cultivating a campus community that is recognized as caring and creative."[28]

There are ideological outliers as well, including those that are more explicitly politically right-wing in orientation, often well-funded by private donors, such as televangelist Pat Robertson's aptly-named Regent University (Virginia Beach, VA) which emphasizes "Christian leadership to change the world" and is known for placing students into positions of governmental influence.[29] Even more elaborately, Wheaton College (Wheaton, IL) lays out an unusually specific guiding credo with a long list of very specific theological touchstones consisting of "biblical doctrine that is consonant with evangelical Christianity," including what is required for salvation (the Lord Jesus Christ died for our sins. . .and "all who believe in Him are justified by His shed blood and forgiven of all their sins"), the "existence of Satan, sin, and evil powers" and much else. Like the social justice orientation of some many of the Catholic universities, these ideological outliers are not all of a politically conservative bent. Though it continues to conceive of its purpose as "to promote the cause of Christ," and "vision of a world shaped by Christian values," Berea College (Berea, KY), one of the first private southern colleges to racially desegregate (until the US Supreme Court forced them to *segregate* in 1908!), pursues a strongly progressive-egalitarian agenda by making a "tuition-free pledge" to students and preferentially admitting those with low incomes.[30] Berea sees this as of a piece with its commitment to racial, gender and other forms of equality "to create a democratic community" and "assert the kinship of all people."[31] There are also several independent progressively-oriented schools like Oberlin College (Oberlin, OH) which does not reference as comprehensive a worldview, but dedicates itself to "enduring commitment to a sustainable and just society" and is known for its very liberal campus culture.[32]

The sociological backdrop of these outliers is that religious belief is sharply declining among young people in the US.[33] However powerful and visible certain religious schools might be, in the American education system as a whole around 90% attend public schools and almost three-quarters of college students attend public institutions.[34] Therefore, given the constitutional prohibition against government establishment of religion in the US, the vast majority of its students inhabit settings explicitly forbidden from embracing religious worldviews which, despite the abovementioned erosion of religious belief and church affiliation, still remain the most common type of comprehensive worldview among those who hold any kind of worldview. This situation raises the question of whether a more fully fleshed-out secular worldview of some kind might be possible to help orient these institutions and play the role filled by the comparatively clearer vision of the religious schools. Secular yet substantive worldviews—the kinds that might generate a paideia in the classical sense—are hard to find, though.

There have been progressive credos attempting to link left-liberal values to larger issues of ultimate meaning and purpose. Notable among these is that of the godfather of progressive education himself, John Dewey. In his strange and seldom-read *A Common Faith*, Dewey attempts to distinguish between traditional religion that is reliant on a concept of the "supernatural" and a more general "religious experience" that is natural or "common" to all human beings.[35] For Dewey religious *experience* as such is a cultural universal and it should be rescued and reclaimed from the superstitions that have been woven into it by traditional religions, Christianity being especially suspect due to its extreme reliance on supra-scientific claims. In place of such atavisms he essentially substitutes engagement in scientific discovery, collective moral progress (away from exclusive tribalism) and a generalized historical-anthropological appreciation that:

> We who now live are parts of a humanity that extends into the remote past. . .The things we most prize are not of ourselves. They exist by grace of the doings and sufferings of the continuous human community in which we are a link. Ours is the responsibility of conserving, transmitting, rectifying

and expanding the heritage of values we have received that those who come after us may receive it more solid and secure, more widely accessible and more generously shared than we have received it. Here are all the elements for a religious faith that shall not be confined to sect, class or race. Such a faith has always been implicitly the common faith of mankind. It remains to make it explicit and militant.[36]

Dewey raises an interesting proposition: can one separate what he takes to be a natural human religious attitude from any accompanying theological trappings? He argues that the dividing line between unjustified and justified "faith" is belief in the supernatural, the positing of extra-terrestrial divine power for explanation, supplication, etc. Against this Dewey indeed suggests attending to something extending beyond one's own individuality: the "continuous human community" that can now function as a naturalized religious belief purged of gods and demons.

While Dewey's writing style may be uninspiring (there are superior secular humanist belle-lettrists), it does express the core of what one might non-disparagingly call a "secular humanism" whose originating self-conception is as deliverance from the mental shackles of the past. Yet it is also as extreme a form of human-centeredness as one can imagine, an apotheosis of anthropocentrism that seems subject to a flaw as deep as that of which it accuses religion: it does not account for its own normative underpinnings. In a way it is subject to a *moral frame problem* analogous to that which bedevils the philosophy of mind, namely, it arbitrarily focuses its moral perception solely on humanity and unaccountably prizes humanity above all else.[37]

But how is this exclusivity to be justified? Is it on account of the match between the object of the inquiry and the inquirer? Why should this matter? Consider how we are a great many things along with being "human." For example we are also mammals, living creatures, earthlings, denizens of the Milky Way galaxy, DNA bearers, concatenations of compounds and molecules, *ad infinitum*. If the presumption is that we must always root for the home team, why not also defend mammalism, biocentrism, Milky Way galacticism, geneticism, molecularism, etc.? It all seems rather arbitrary to single out "humanity" as the primary object of moral concern. At least the Christianity Dewey detests as antiquated has a strong concept of *humility* that militates against the infantile tendency always to place ourselves at the center of everything. If a key to finding meaning is the ability to place oneself vis-à-vis that which is larger, *ipso facto* it is also necessary to maintain the distinctiveness of that larger entity. (Marx incidentally commits the identical chauvinistic error in his recourse to the notion of a "species being," a bit of biologistic backsliding from Kant's comparatively capacious moral inclusion of all "rational beings.") Inflating oneself into the sole source of value and the center of the moral universe seems cosmically infantile and unsatisfying.

In addition to the framing problem, even on its own terms Dewey's formulation aims too small. Identifying with a transhistorical humanity has an ostensible appeal and it may be thought of as an extension of the egalitarian impulse ("you are me" and "I am you" and "they are us" and so on). But on a large enough cosmic timescale, even the entire history of humanity is bound to be an almost infinitesimally small blip. Do the math. No matter how much adaptability and resilience we demonstrate in the coming years, no matter how successfully we might escape Earth to terraform other planets and no matter how completely we might download ourselves onto disparate substances, it will all come to an end, finally, at some point. Infinity is a long, long time. The only way out of this problem of long-term demise would be to posit some kind of *a priori* cosmic assurance of human continuance, a species immortality guarantee that would be equally implausible as the individual immortality guarantees of the very biblical religions supposedly being advanced upon. Oddly, as zealously as Dewey critiques the supernatural, he also refuses to embrace the *natural*; ironically, he thereby conducts his own leap of faith through an arbitrarily selective application of moral concern. Dewey's proposed common faith has sensible elements but ultimately it falls short.

The ideas behind Dewey's *A Common Faith* contain the essential core of the secular humanist approach and despite the above criticisms at least they strain for a larger narrative and sense of meaning. As such they are light years ahead of the empty marketing self-presentations currently characteristic of most American universities. One can even imagine a salutary Deweyan makeover of many of these statements where at least the key ideational dots would be connected up into coherence. Could such an imagined politically progressive vision be made to reverse the centrifugal force of the identity dissensus that now threatens traditional egalitarian principles like free speech, due process and, ironically, equal protection? The American right-wing has convinced itself that most universities, certainly the most elite ones, have been wholly taken over by the villainous "cultural Marxist" *bêtes noires* of their imaginations.[38] The error here, however, is to assume that the identity mongering of the "social justice warriors," those whose primary mode of activism consists of online call-outs of the benighted and showy "anti-platforming" tantrums over campus speakers with whom they disagree, is coextensive with the political left as a whole. In the overall scheme of American politics, the left may not amount to much, but thankfully there is more to it than the collegiate call-out crews.

There are substantial and growing segments of the left that adhere to the more universalist Enlightenment-based egalitarianism that has traditionally served as the moral basis for progressive political projects. Even if they don't realize it, this non-identitarian left is closer to Dewey than they are to the incipiently authoritarian exclusivity of the identity obsessed. So although there are strategic overlaps on particular issues, most campus-based identity partisans bear little ultimate resemblance to actual socialists. The largest socialist organization in the US, the Democratic Socialists of America, operates under a humanist and universalist platform whose very first item of belief is that "both the economy and society should be run democratically to *meet human needs*, not to make profits for a few."[39] For the egalitarian left, in the best case the different strands of identity politics can be gathered up into some kind of solidarity, presumably via a political education process that deepens individuals' critiques and illuminates the interrelations among the ostensibly disparate struggles. However, there are less sanguinary prospects that are beginning to become apparent, as identitarians seem mostly adept at cannibalizing their (often) strategic left-liberal allies while simultaneously strengthening and emboldening the right wingers they claim as mortal enemies. Generally speaking, the right *loves* these kinds of fights and makes no secret of it. As former Trump adviser and Breitbart chairman Steve Bannon notoriously stated, "The longer they talk about identity politics, I got 'em. . .I want them to talk about racism every day. If the left is focused on race and identity, and we go with economic nationalism, we can crush the Democrats."[40]

In this respect one imagines an American political version of the "Old Firm" as the Scottish soccer rivalry between the Glasgow teams Celtic (Catholic) and Rangers (Protestant) is somewhat cynically called. The two sides are hated rivals and the history of their derby is replete with incident yet it is clear also that they require one another and—certainly from a business standpoint— are in a kind of symbiotic relationship with one another.[41] As is always the case with identity formations, the sharpness and definition of their own identities is in large part shaped by their opponents, their Other, in this case their perennial crosstown rivals. Similarly, identitarians *need* their Milo Yiannopouloses, their Lauren Southerns and Brittany Pettibones and, better yet, their (actual) neo-Nazi Richard Spencers. Their on-campus visits provide powerful moments of consolidation for them and their beloved stark borderlines between those "with us" and those "against us." If they were honest they would admit how ecstatic they are when they have such palpable opposition. Analogously, I have been at many stalled protests that were "saved" by police intervention, that "now we see the violence inherent in the system!" moment.[42] Though of course mutually contemptuous, these two (alt-right/alt-light vs. antifa/identitarians) seem locked into a mutually reinforcing kabuki dance where the outcome has mostly to do with the public poses they are able to strike to their own referent groups. In a sense the primary objective seems to be augmenting their own brand loyalty above all else.

Thus the energies released by these identity-fueled symbolic confrontations flow like hot liquid metal into the consumerist procrustean smelting beds laid down by neoliberalism and its individualist ethos to trap emergent alternative energies. As one runs this particular tape, the predetermined endgame is that everyone ends up sentencing themselves to an identity-island. The intersectionality of some of the inter-island waterways may create some salubrious inter-connections but overall the islands keep getting smaller and smaller, more and more exclusive as, reality show-like, participants get kicked off the collective island due to insufficient purity. It is a final vision out of the Werner Herzog film *Aguirre the Wrath of God*, where the zealous and authoritarian conquistador protagonist (played by Klaus Kinski), everyone else departed or dead, is left shouting maniacally and decrying his fate all alone, with nothing left to him but to float downstream by himself on his raft.[43] The undeniable telos is toward wholly individualized sinking islands that can remain pure because they are all one of a kind, an ironically scaled mass solipsism prefigured by trends in our consumption habits, a species of the "bowling alone" phenomenon made famous by sociologist Robert Putnam or the internet-enabled filter bubbles through which information travels.[44] The faux togetherness of its imagined communities exposed, identity politics thus reveals itself in its loneliness and self-regard to be, in Adolph Reed Jr's phrase, "the left-wing of neoliberalism."[45]

Universities and even their internal critics are being hollowed out by capital and by a continuing market fundamentalist streamlining. As in Richard Russo's comic campus novel *Straight Man*, students have un-ironically become "customers," scholarship functions as a revenue stream and university presidents are essentially marketers- and fundraisers-in-chief, etc.[46] There is no transcendent purpose of meaning to animate anyone's efforts, no grand narrative beyond increasing alignment with corporate sensibilities and more efficient servicing of consumerist selves. There remains but a ghost of public service within the shriveling corpse of the land grant university that was once animated by those ideals at its nineteenth-century inception. This structural nihilism is perhaps the deepest problem afflicting contemporary educational institutions. It is especially apparent in higher education because of that sector's market sensitivity where students are customers who pay—and pay through the nose—for this investment property. Originally it was a triumph of inclusion and *modus vivendi* liberalism: religious and business groups (and others) coming together like intersecting sets in a Venn diagram to form a common school project.

The non-paideutic and non-eudemonistic nature of contemporary public education was at first a virtue that, along with other institutions like the military, allowed a heterogeneous society to develop some sense of a shared identity. But at this point the only possible mission statement for a public university (or public school district for that matter) is so platitudinous that it can have no determinate meaning; it just doesn't say *anything*. It's just *bullshit rhetoric* in precisely the sense that philosopher

Harry Frankfurt (2006) meant it: "the bullshitter doesn't care if what they say is true or false, but rather only cares whether or not their listener is persuaded."47 They are all things to all people and profess to represent anodyne "ideals" to which nobody could object which, of course, means they are pursuing nothing in particular—except possibly their own growth and enrichment. They have come to stand for nothing. And despite the rhetorical noise they generate via the virtue smoke signals of campus politics, they produce no new visions and have no resonance outside their own peaceable little kingdoms. Nobody looks anymore toward the marketing conglomerates known as "universities" for wisdom and purpose. One might as well look to Exxon-Mobil or Apple or Goldman Sachs.

Internally, this purpose void has been filled by the individual identity projects of the constituent members of the university community, especially the younger generation, the customers, who are in the most churn in the liquid modernity identity blender. So the vacuum of substantive leadership and the campus call-out culture seem perfectly paired with the bottom-line brutalism toward which today's universities have been trending. Bereft of any larger political vision, this campus call-out and de-platforming activism spins out into therapeutic and careerist economism, where the remedies envisioned have to do with such as the creation of new academic and corporate niches in HR-type managerial positions to oversee the inclusion where, as literary critic Walter Benn Michaels suggests, "the model of social justice is that the rich make whatever they make, but an appropriate percentage of them are minorities or women."48 A slight diversification of our credentialed overlords will allow us all to sleep well. It seems that identitarianism leads us into two equally undesirable destinations: either the petty authoritarianism of rejectionist authenticity cults (mirroring their right-wing counterparts) or the maintenance of a managerialist strata of institutionalized diversity bureaucrats.

If, depressingly, the education system—and even its own campus critics—looks too sunken into neoliberal nihilism to provide inspiration, where might one look? Traditionally, politics is of the major available avenues through which to pursue the good. Might then the universalist political left *as a whole* do any better at providing some positive normative vision? It seems worth a look because they do seem convinced of something. But what exactly is that something?

Chapter 3

The hollow left

Victory for the Forces of Democratic Freedom!
David Foster Wallace (2000)[1]

Moral superiority?

Nobody admits it but most on the left feel morally superior to their political adversaries. They are better people; the *correct* kind of people (and I remember when "politically correct" was used un-ironically on the left; it was jiu-jitsued into a slur by the right, not manufactured by them). Yet one cringes at the phrase "morally superior" and both of its component terms. "Moral" still carries the connotation of the Victorian scold keeping up bourgeois appearances. And "superior" implies a trait that irony-imbued *Homo sapiens hipsterus* has long tried to eliminate from itself: earnest self-righteousness. Moral superiority seems an anachronism from a bygone era of unreflective Church Lady religiosity, currently held onto by holdover true believers in dismal parts of the country best left unvisited.

But the common veneer of cool relativism is obvious bullshit. *Everyone* makes comparative moral judgments whose objects are both individuals and groups, where we allow generalizations about *those people*. All of us constantly do this and it probably can't be helped due to our primate hard wiring. Nonetheless, in a postmodern age whose hallmark is what the philosopher Jean-Francois Lyotard called "an incredulity toward metanarratives," it is extremely uncomfortable to admit this persistence of judgmentalism, which seems to belong to a bygone era when there was greater unanimity concerning worldviews. Without that shared framework, it is hard to make value judgments with any level of assurance. With it, one could more confidently denounce so-and-so's wretched behavior as, say, "un-Christian" because everyone around understood the meaning of the accusation and shared at least a vague background notion of what we as human beings were supposed to be doing. In the developed world, however, the previously-dominant metanarratives that once licensed all the moralizing lie shattered in a thousand pieces.

What remains is the basic impulse to judge, but minus the shared comprehensive conception of the good that previously gave the execution of that impulse social confidence and ideological coherence. Bits and pieces of the ancient tradition remain, though. Despite having been strained through a secular membrane, a successor moral schema upon which the left's moral judgments are made may still be detected. It is not as easy as seeing how the right bases its own moral judgments, which— however self-deluded—spring from traditional religion and also bourgeois notions of thrift and self-reliance, the latter often lip-serving a mythical "rugged individualism." But it is there; it is a bit harder to detect, but progressive types still judge on the basis of *something*.

The contemporary left seems particularly invested in an odd species of quasi-perfectionism, "perfectionism" being a philosophical term for a worldview that erects an ideal type of human being to serve as everyone's regulative ideal. Though it is a vague sort of perfectionism, those on the egalitarian left end of the spectrum seem especially prone to seeing—and perhaps *needing* to see—themselves as models of *better* people than those they are opposing on the right. This "better than" is not quite coherent but it implies as a basis of judgment an image of an ideal person that covertly premises much of the left's characteristic attitudes and moral postures. This person is first and foremost radically egalitarian, not just in terms of formal public policies, but in terms of a basic attitude toward life. *All* human beings (and maybe sentient beings) should be treated with an equivalent quantity of moral concern, no matter their identity or origin and regardless of geographical distance.

Correlatively, a vigilance is required to police deviations from this norm of equal treatment regarding both atavistic prejudices (racism, sexism) and emergent ones against newly acknowledged identities such as the transgendered and those constituting the wellspring of the "+" in LGBTQQIP2SAA+ LGBTQQICAPF2K+. The "K" recently under debate is for "kink," representing, for example, BDSM lifestylers, who undeniably face prejudice and deleterious consequences from being "outed," etc.[2] As potential objects of persecution, members of the "kink community" share an important defining trait with members of oppressed groups with more traditional identities—however reluctant members of the latter may be to extend solidarity. A more studied but analogous controversy involves the apparently relatively low support for same-sex marriage among African-Americans.[3] (Despite the phenomenon of intersectionality, membership of a traditionally oppressed group is no guarantee that one will be more alacritous in one's sympathies.) The potential for anti-kink discrimination came to light in the case of the Canadian Broadcasting Corporation's Jian Ghomeshi, who was fired from his position as a television morning show host allegedly purely because of a negative perception of his BDSM lifestyle.[4]

Whatever the approved targets, leftist denunciations and shaming (typically online via social media) usually have to do with perceived lapses in the required level of personal moral capaciousness, where some kind of disregard or animosity is detected in an offender's actions or words. This is what sin has become for the contemporary left: an *egalitarian deficit disorder* visible by word or deed, qua perpetual evidence of personal fallenness needful of a neo-Maoist digital struggle session. Like its Christian forbears, it has even developed what journalist and alt-right expert Angela Nagle describes as a bizarrely "self-flagellating politics," a digital hair shirt wearing, designed less to purify the miscreant than to abase the rest of us who have not yet come clean or, one is tempted to say, "gone clear" as in the confessional ideal of scientology.[5] Nagle elaborates this impulse: "Publicly declaring your sins makes you appear a better person than those who have not declared them. It is not really a put-down of oneself, but a put-down of others, who are less morally worthy for having been less forthcoming in their confessions."[6] And indeed, these public confessional morality plays grow more and more elaborate. Unlike the older self-abasements, however, they are strangely detached from religion or, odder still, any other comprehensive conception of the good life for a human being. In this sense the analogy to the brutal Maoist struggle sessions of the cultural revolution period is limited as well. At most there seems to be a kind of staged narcissism — composed of the social media click-currency of "likes" and "followers" — beneath it all, a mere matter of popularity or fashion that is vestigially invested with an aura of moral seriousness.

The moral pretense is even more apparent when the denunciative energy is externally directed toward the sinful Other. Here the performative self-purification is conducted via an escalating enthusiasm for "naming and shaming" the Other, where the twitter mob is the iconic phenomenon of the moment, functioning as a digital re-creation of the widespread ancient phenomenon of expiating communal sins via human sacrifice. The denunciation also performs the age-old function of identity consolidation — an old trick perfected by tribalists and racists and nationalists — by more sharply drawing the borders between "in" and "out" groups. This was especially apparent after the Trump election, when centrist Democrats could understand Trump's popularity only in terms of opprobrious mass character flaws having to do with "racism," "xenophobia," "misogyny," and the like. The Trump supporter is very clearly portrayed as suffering from a significant character deficit even, really, as *sinful*, an army of evil idiots in red trucker caps *out there* in the hinterlands. Note here also how, under a more cognitive capitalism, "idiot" has become a grave *moral* insult and not just a descriptive label about ignorance or lack of cognitive capacity. The losers in the neoliberal sweepstakes must always appear *blameworthy* (and hence punishable) rather than mere pitiable victims who know not what they do. Given how much the neoliberal winners are invested in the implicit claim that their alleged intellectual superiority is also a moral superiority, "stupid" as an epithet must always be understood as "stupid asshole," which is, incidentally, precisely how flyover Trump supporters *must* be seen.[7]

Admittedly, those on the right tend to view leftists as moral degenerates as well, though the reference points are less Hitler than perhaps Sodom and Gomorrah. On the American religious right, however, no matter how fire and brimstone the sermonizing, everyone is at least in principle *redeemable*; even the greatest sinner can through an act of his own will be saved. Souls can still in principle rise above all earthly determinations, including the most sinful pasts. It is for example, notably easier for them to embrace Trump himself who even supporters cannot see as any paragon of personal virtue. This embrace of Trump by the religious right is often seen by outsiders as mere political expedience and even hypocrisy—and it undoubtedly is to an extent. But it is important to understand how their very belief structures make it easier for some religious types like "born again" evangelicals, ironically, to forgive and embrace *certain types* of sinners and deviants than the left is typically able to.[8] In their own way they are actually more inclusive in that they tend to be less condemnatory of imperfect personal histories so long as the currently-professed beliefs are up to snuff. If *per impossibile* Bernie Sanders or Barack Obama publicly declared their allegiance to fundamentalism and asked for forgiveness for their past beliefs—and they seemed sincere—I have no doubt that the right would eventually embrace them as souls they have won, probably quite enthusiastically. And the right-wing commentariat has long been replete with ex-left-wingers but not the reverse, from Irving Kristol and Whitaker Chambers to David Horowitz and Dave Rubin[9] and in some sense Ronald Reagan himself.[10] There are of course examples going the other way (Michael Lind, Elizabeth Warren and, given her self-description as a "Goldwater Girl," Hillary Clinton come to mind)[11] but these are far less numerous and they are not far left figures. Now imagine, again *per impossibile,* Donald Trump or Ted Cruz becoming, somehow, born again progressives in the mold of, say, Alabama Governor George Wallace's later years and it is impossible to imagine that the now-ascendant identitarian left would ever — *ever* — accept them.[12]

Today's identity-focused left is in principle less forgiving because its conception of moral failure is much more *existential* than it is belief-based; it's much more a matter of *who* you are (identity) than *what* you believe.

The left's supporting social theories are more structuralist, those guilty of racism or sexism are more sociologically *determined* to be that way; in a way they *have to be* racist and sexist because of the locked-in facticity that has constructed their very whiteness and maleness. To abandon this conviction would be to abandon any analysis of the structural dynamics of oppression. Partly this picture arises from the left's wholesale embrace of Antonio Gramsci's critique that structural oppression is to be found not just in economics but also in the cultural sphere. (This was originally a device for explaining why workers in the West never revolted as they were supposed to as per classical Marxism.) This cultural oppression is so powerful that it causes workers to abjure economic self-interest, to the extent that the cultural sphere comes to be seen as nearly an all-determining source of political motivation. "What is the matter with Kansas?" is the same thing that was the matter with Gramsci's Italian factory workers in the 1920s, Germans after World War I, Americans during the Depression and so on. People are reconciled with their lot—even to the point of going against their own material interests—because they are wholly besotted with hegemonic cultural-ideological distractions having to do with religion, race, nation, etc. This delays the revolution.

Thus the basic Marxist theoretical framework and its long-term optimism is preserved. But the Gramscian analysis comes at a high epistemological cost: the reliability of workers' "consciousness" is now thrown into question indefinitely. The working class—even the whole lot of them—can now be *mistaken* about what they think they want and, importantly, "populism" is therefore not necessarily to be trusted, a conviction further reinforced by ensuing fascist movements. Though the left is always *in principle* on the side of "the people" it is now more than possible for the left to need to position itself against the *actually existing* people temporarily. Left politics still exists within this temporal gap. As such those toward the left end of the spectrum too often must position themselves as moral educators of the people, who are not yet ready to be trusted. They are thus unsurprisingly rejected by many of the members of "the people" as, at a fundamental level, disrespecting of them as human beings. Predictably, right-wing populism steps in to fill this correctly-perceived respect vacuum, promising, for example, not to tutor them about how to talk.

Due in part to generations-old theoretical commitments, vis-à-vis middle America the left has dug itself quite a hole here. And it keeps on digging. Due to the identitarian fusion of structuralism and the Gramscian theory of culture, the left has moved beyond its traditional posture of *educating* ordinary people to one of *condemning* them as wicked. In addition, again courtesy of the aforesaid theoretical commitments, whereas the religious right can often "love the sinner, hate the sin," the left is typically required to hate *both* of them. "Checking your privilege" is not like checking your coat at the cloakroom, where you then proceed to enter sans coat. There is no implication that white or male privilege can actually be *removed*; it can merely be "called out" as an object of shame. It is an existential condition that can be acknowledged but not really altered and so there is no avenue of participation for the alleged privileged one—save perhaps for some sort of Maoist spectacle of self-abasement. In fact, claiming to have successfully removed white privilege, for example the assertion of a thoroughgoing "colorblindness," would *itself* be symptomatic of privilege, even its very paradigm, i.e., the luxury of being able to "pass" through the societal mainstream without a racialized or gendered identity.

In a strange twist, the centrist portion of the left-liberal spectrum has actually grown *more* strident regarding cultural politics, an outcome that one can see in retrospect was determined by liberals' historical dismissal of class politics during the neoliberal ascendancy starting in the 1970s. By the 2016 election, Clintonian liberals seemed to be the true heirs to Gramsci, with most of the post-mortem analysis pointing toward the cultural bamboozlement of the white working class, from their entrancement with celebrities and reality TV to their various -isms. (And of course Russia did it.) The far left—from Sanders leftward—at least recognizes the absence of a class critique of the American economy and so tends to stop short of outright contempt for middle America. However, there is still the structuralist tendency across the whole left spectrum to see politically-relevant identities as fixed and essentialized, and consequently also the attitudes and behaviors associated with those identities.

Currently under the aegis of elements of Black Lives Matter and other activist groups, the far left plays into this dynamic: your racism is ultimately attributable to your unavoidable white privilege which itself is attributable to your unavoidable sociological ensconcement in a white supremacist hierarchy. And if *per impossibile* you claim to "rise above it" then that very conceit that you have so risen is itself further proof of your inescapable infirmity. Since they cannot be differentiated, sinner and sin become one and the same and are *both*, therefore, eminently hateable. With an historically ironic Calvinist twist, "deplorability" is largely predetermined. One might even go full Weber here: anxiety among the lovables over the possibility of still-lurking inner racist demons may perhaps be thought of as a concern that residual sin will blow one's cover as an "ally." If the Brooklynite or Portlander discovers upon introspection that she harbors a racist idea, what is far more horrible than the sin itself is that its existence may be symptomatic of her own deficient moral status. The sin is a *sign*. As per Weber's "Protestant ethos," the Calvinists-cum-Puritans were *the* original virtue-signalers, seeking to reassure themselves as to their status as one of God's predetermined elect.

This may be one reason for the left's continuing affinity for protest marches, which have largely been ineffectual since the civil rights era: they allow for a public display of virtue that is inherently rewarding, a reassurance, in the old Calvinist sense, that the participant is to be numbered among the saved. At best, the implied rationale that mass public marches and the like "raise consciousness" and educate the public about particular issues, though held to as an article of faith, remains highly contestable. On balance, the collective demonstration of virtue may repel more external observers than it compels. Who knows? That no one bothers really to find out (beyond crowd estimates or a TV-ratings style follow-up poll) is itself telling; the beloved demo is, ironically despite its public execution, *inwardly* directed as an identity consolidation among participants, an exercise in psychological reassurance. Civil rights marches like Selma were different in that organizers' aims were to create a spectacle where they exposed themselves to danger to win over a larger public. In-group cohesion undoubtedly occurred but this was a byproduct not a main goal.

The impurities of conscience thus raise an acute problem, one of the oldest in Marxism. Despite the more or less structuralist assumptions about economics and history, how is it possible for one group to gain enlightenment—critical intellectuals, party vanguard, the left in general—over and against everyone else who remains, *ex hypothesi*, pinned like butterflies to the mentalities appropriate to their prescribed social positions? As per Marxist materialism, where ideas are essentially epiphenomena of material conditions, the bourgeois is supposed to think only bourgeois thoughts, the proletarian, proletarian thoughts and so on. Marxism solves the problem readily enough for the working class themselves who, again *ex hypothesi*, have their identity, character and beliefs forged for them through their own lived experience as class members, in effect playing their role just like everyone else. (In many respects one should be no more angered by the greedy capitalist than by the actor merely playing a greedy capitalist in a show; we all must play out our roles.) There is nothing incoherent about this conception of everybody playing their assigned roles and having those roles' attendant ideas. The conceptual problem arises when a cross-class interloper, say Marx himself, claims to be able to rise above it all and give everyone a bird's-eye view of the whole. For how does the analyst *him or herself* rise above the push and pull of class antagonisms in order to witness a categorical "truth" of the situation that is not merely an artifact of his own class situatedness? Where does the intellectual and would-be truth-teller stand?

There is an analogous problem among today's identitarians involving similarly interloping cross-group figures such as the antiracist white person and the feminist male. Zeroing right in on this ideological weakness, this is the figure pejoratively termed the "cuck" by the alt-right, a label that is purposefully offensive but whose staying power in the twittersphere is attributable to the grain of truth the sophomoric (and authentically sexist) accusation contains: that people who act as if they are so noble as to cast aside self-interest are full of it. It is a juvenile calling out of the pretense of some that they are such moral saints that they have wholly transcended self-interest. It is a refusal to countenance the public displays of self-righteous virtue-signaling and self-flagellation. In this respect the cuck insult actually hits an uncomfortable place on the left, the special theoretical exemption many on the left give to *themselves* from the undignified social determinism with which they routinely dismiss everyone else's mentalities. Turnabout is fair play.

Anticipating this problem, Gramsci himself had tried to square this circle with the notion of the "organic intellectual," a kind of semi-autodidact who is supposed to arise to her leadership position from *within* the working class and thus remain identified with it regardless of her flights of mind. It does not seem that the contemporary identity-based left has anything analogous. The closest attempt at a bridging hybrid type seems to be an implicit *noblesse oblige* on the part of rich corporations and elite universities to diversity train their members, the professional diversity trainers being the latter-day vanguard, the commissars of corporate inclusion, working to legitimate inequality by ensuring that the 1% "looks like America." More benignly, at the moment outside of officialdom, Black Lives Matter probably provides a more authentic example of the "organic intellectual" notion with its emphasis on internally developing leadership through local organizing.

The egalitarian left

Before proceeding with this discussion, there is one historically influential attitude toward this entire question on the left that should be addressed. It is a kind of deflationary attitude that reflects optimism about the "science" part of "social science," from a century or two ago when such sanguinity seemed more justified.

At least since the later writings of Marx and Engels, there has on the left existed a school of thought that denies the question of morality altogether. The idea here is that what is being described in leftist critique is merely descriptive in the manner of *science*. It's not that the working class *should* make revolution in any *approbative* sense, that one should be in favor of it because it is *good*. Rather, because of the dialectic or historical materialism or whatever methodology, the oppressed will gain class-consciousness—and ultimately species-consciousness—because it is the *inevitable* outcome of causal processes having to do with the dynamics of capitalism, etc. From this point of view it would make no more sense to defend any *moral* imperative pertaining to socialist revolution than it would be to ascribe the advent of Spring to the Earth's morally good intentions. It will simply *occur* for causal reasons pertaining to science. (Strictly speaking, from this "scientific socialism" point of view *blaming* capitalists—in the sense of moral opprobrium—is also wrongheaded.) In certain nineteenth-century circles there was once great optimism about how scientific, and hence predictive, the "human sciences" (e.g., anthropology, history, politics) could become. Suffice to say, however, there is decidedly less so today. There is widespread appreciation that these kinds of analyses are not analogous to the hard science as, to put it briefly, their "data" are so much more thoroughly open to interpretation). It has become axiomatic, in fact, that disciplines such as history, while they have their own recognized methodologies and quality measures, are very little akin to the hard sciences like physics, chemistry or biology. They pitifully lack predictive power, for one thing. And all historians recognize the value of new histories of previously-written about phenomena—no matter how competent was the previous scholarship—because of the need to examine the phenomenon through the lenses and with the concerns of the new generation. For such reasons it seems safe to dispense with those who claim that their politics are scientific and as such are impervious to analysis of their explicit or implicit moral claims. *Pace* Marx and Engels, then, no political view is above the analysis of its moral implications.

Clearing away the scientific inevitability option allows the discussion to begin.

I'll try to define "the left" and its core views as concisely as possible. Using the frame of American politics, I define the left as those tending to emphasize "equality" more than those on the right who tend to favor "liberty." This leaves out a lot (like foreign policy), of course, but it is the most economical way to differentiate left from right that I can manage. From liberals to the farther left, universalist principles of egalitarianism, equal protection, tolerance, inclusion and full participation are the signal norms. Not necessarily the fever dreams of far leftists, but the marquee left-liberal accomplishments in US history have been mostly along these lines: religious disestablishment (via the First Amendment), the abolition of slavery, racial desegregation (via the Equal Protection clause), voting rights, extension of the franchise and the rights of women generally, disabilities rights, access to the legal system (e.g., a right to counsel for indigent criminal defendants), etc. This is hardly exhaustive, of course, but it illustrates the general trajectory of the left-equality side of the policy ledger.

Even as the antagonism deepens between former Hillary and Bernie supporters, the broad left-liberal spectrum does have some common characteristic allegiances that can be summed up in a word: *egalitarianism*. In fact, a simple way to put the major current divide has to do with the extent of egalitarianism in the economic realm, where the central charge against Obama-Clinton neoliberals by socialist-leaning Sanders types is precisely that the liberals are unwilling to extend their historic egalitarian commitments to *economics*. The 2016 Democratic Primary flash points were mainly along these lines, such as the largely quantitative distinctions between the two sides on college tuition, minimum wage and universal health care (e.g., the far left single payer plan possessing a greater universality than the insurance-driven "Obamacare"). An exception being a distinction over globalism and trade, where equally fatuous proposals were being advanced: Clinton still clinging against all evidence to Obama's neoliberal magical realism concerning the benefits of globalized free trade and Sanders, like Trump, implicitly promising a reversal of de-industrialization to "bring back jobs." For present purposes, then, the actually existing left may be defined as a set of political views where the commitment to egalitarianism—and related concerns such as inclusion—is the core motivator of policy proposals. To the extent that one allows egalitarian imperatives to pervade the economic realm, one is "far left," whereas one who limits egalitarian sentiments to matters of inclusion such as formal civil rights and identity politics, is at the more centrist/liberal end of the spectrum. The far left cannot forgive the economic blindness of the liberals while the liberals cannot tolerate the far left's heretical refusal of what Mark Fisher termed "capitalist realism" as they strain to imagine alternatives to neoliberalism.[13] However they stretch it, though, what both ends of the left-liberal spectrum share is an egalitarian inclusionism.

Beneath any accidental cultural snobbery, this egalitarianism, I suggest, is what accounts for most of the left's sense of moral superiority. A crude recent example of this sentiment is the centrist kneejerk response to the election that saw Trump voters as atavistically racist, xenophobic, misogynous, homophobic and anti-everything else. (Never mind that Trump voters included many previous Obama voters, a majority of white women, a majority of married women, nearly a third of Latinos and 15% of African-Americans, etc.) Raw benighted prejudice and gloating white supremacy allegedly accounted for Trump's margin of victory. This analysis was especially pronounced among centrist Hillary supporters who, true to form, retreated into their identity politics redoubts and were mostly contemptuous of analyses that tied Trump voting to economics and class issues. From this perspective the explanation for the electoral catastrophe is comfortingly simple: the people in the blue coastal strongholds, who just happen to be the ones situated in the updraft of neoliberal globalism, are simply less racist and small-minded than their flyover compatriots. Their personal virtue has simply evolved more felicitously and their morally capacious way of seeing the world is worthy of comparative approbation because their egalitarian commitments have awakened them to a better worldview and way of living. In this respect their virtue-signaling is not cynical but rather has become internalized; this is their self-understanding. Personal growth is like an existential quantifier in logic that expands the circumference of their moral regard from "some" to "all" people on the model of 1950s and 1960s civil rights struggles.

Moral virtue is here conceived as neatly linear and "progressive"; virtue is proportional to one's ability to include *all* and consequently vice is measured by how short of this universalism one falls. This is a teleological view containing a strong belief in progress, or at least continued progress's desirability. This triumphalism draws inspiration from the history of formal civil rights such as the overcoming of voting rights exclusions from the property-less to African-Americans to women, and so on. As with capitalist production itself, it is a growth model that must perpetually expand, otherwise the inclusiveness is experienced as having been halted. So there is a zealous attempt to locate newcomer identities for an inclusion process whose very continuance provides the requisite reassurance that moral progress is taking place; there must always be a frontier that is being breached, a ceiling being shattered. Any retraction or retrenchment can therefore only be seen as anti-progress, a violation of almost naturalized expectations of forward momentum.

Yet however large the moral balloon, in order to have any kind of reality, it is still constituted by a membrane—a border—

that needs to be affixed in some manner; even as newcomers gain insider status, others must be ejected to become outsiders— otherwise the thing loses definition. It is this perpetual need for differentiation that is obscured in the left-liberal model of political morality as perpetual expansion. Moral rejects are needed just as much as are admittees.

Moral ejection is typically perpetrated for one of two reasons: 1) style/aesthetics and 2) strong non-secular beliefs. To be sure, there are some outright racists and misogynists and phobics of various types who are often excluded, too, although ironically these exclusions are not wholly consistent, for example, it is forgivable to call young black men "superpredators" for political gain, whereas questioning a birth certificate for political gain is obviously and irredeemably racist. Another example is how anti-semitism and misogyny are frequently forgiven as cultural tolerance if they are of Islamic origin. The style component, otherwise known as raw class snobbery, is underappreciated. Having lived and worked among elites my entire life, I can attest to the visceral reaction against mostly poor people (unless they're the admirable kind of poor such as aspiring artists, musicians, writers or humanities majors), not in the form of being "anti-poor," but carried out in terms of enforcing distinctions of taste in cuisine, film, dress, exercise habits, language, personal hygiene and the like.[14] Abjuring someone who is fat or who wears the wrong jeans or who is not caught up on corporate and academia's latest term of art for a minority group (e.g., though the phrases are formally equivalent, saying "colored people" rather than "people of color" clearly signals moral degeneracy). Regardless of any content the mere *sight* of a Trump rally triggers many a coastal into condemnation of "the idiocracy": bad hats, stupid music, WWE-like production cues, ubiquitous tacky Americana, and just the "look" of the people in the crowds. The ensemble causes a reflexive revulsion among the more cosmopolitan set.

Perhaps more substantially there is also among liberal elites a

127

widespread disbelief that people could really believe the stupid religious shit that seems so pervasive in these morally no-go areas of the country. These people never received the meme about "incredulity toward metanarratives." Interestingly, evangelical Christianity has its own kind of universality, having to with each soul being in principle savable, and its own expansive, proselytizing momentum that, embarrassingly enough for the progress narrative, is proving to be quite a bit more convincing than secularism to large populations in Africa and Latin America. Structurally, this evangelical version of moral universalism strongly resembles classical liberalism's own ambitions. But the religious proselytizing kind of universalism does not "count" because the actual contents of the belief system are an anti-modern mythos that is essentially fantastical and as such is not appealing to anyone's rational capacities. It is not often openly admitted (though it is privately), but most well-educated left-liberals regard evangelical-style belief as some species of mass mental illness, just one step away from Jonestown Kool-Aid. If one jettisons baseline rationality, then one could end up literally anywhere as far as worldviews.

For its part, the farther left believes much the same but adds even greater sanctimony by accusing more mainstream liberals of ignoring socio-economic class. For the tiny sliver of actual Marxists this definition is all-important and is not equivalent simply to "poor people" but rather has to do with a group's position as exploited labor within the capitalist production process. The far left sees itself as somehow able to perceive class dynamics in a way others cannot, partly because of a will to truth for analytic precision (traces of the pretense of providing a scientific socialism). But their perceptual acuity is not simply owing to their big brains; it is not merely a more *accurate* vision. It is a *morally* better one as well because that perception is made possible because it is undistorted by elites' economic domination and other predations—it has not *sold out*. It too is

Is this not defensible? E.G. Reed aanis the PMC or an interest bound w/ liberalism?

like the evangelicals in that its message is possible because it is less occluded by *sin*. In the election aftermath, Sanders leftists have often critiqued Clinton liberals for the abovementioned overreliance on identity politics and snobbery toward flyovers — and there is such tempting low-hanging fruit, such as prominent liberal blogger *Daily Kos* urging readers to "be happy for coal miners losing their health insurance," i.e., hoping Trump-supporting miners in eastern Kentucky now get lung disease[15] — yet the form of the critique is basically the same: decrying the benightedness of excluding a certain demographic (viz., white working class) from the charmed circle of moral concern. They are out-liberaling the liberals in this regard, calling them out on their own callings out. And despite their pretenses to greater universality (i.e., including class dynamics), like Marx himself, the far left hereby embraces a troubling performative contradiction having to do with that perpetual question of how *they* — this small minority of citizens — are able to see with greater moral clarity even than their liberal cousins.

Why indeed? Logically, and again excluding the hypothesis that it is all a dispassionate debate about the objective accuracy of one's theoretical diagnosis, it must be that this is a small morally elite group — a vanguard if you will — who for whatever reason is just *better* than everyone else as demonstrated by their perspicacity. These are people whose moral perceptual apparatus *includes* people that the others are failing to see. As such, they see themselves as *literally* "the light of the world," exemplary moral beings whose personal qualities "giveth light unto all who are in the house [Matthew 5:14-16]." They somehow possess an empathic capacity that others lack to illuminate human needs; to wax philosophical, they grasp the universal while others around them — the discriminators, the coercers, the –ists and –phobes of all descriptions — are only able to see the particular. They are our moral teachers and, when necessary, our righteous tribunal of judgment. This is, to say the least, an overly flattering implied self-image.

Moral superiority can still exist, however. But it is unreasonable for any one political grouping to think that they *alone* possess it. What is ultimately necessary are not better *people* but the building of better *worlds* for us wretched sinners—as we actually are, here and now—to inhabit. This is precisely the educational project the left has abandoned since communism: *world building?* It has instead ceded that task—public education in the grand sense— to atavistic fundamentalisms worldwide (nationalist, Christian, Islamic) and left its own moral judgments blowing in the wind, untethered to any *eudaimonia*, any overall picture of the good life for human beings, any deep background against which to make sense of life's grander purposes. Somewhere along the line we became too cynical and ironic, too ecumenical and world weary, to sustain such an enterprise. No more pie-in-the-sky, no more bloody utopias.

But this "realism" turns out to have been a world historical mistake. The right has understood far better how overrated is the "reality" of a given moment and as a result their messaging is winning. "Occupations" and protest marches alone will not even come close to standing against them. Note how the right has come to power precisely *without* mass demonstrations and the left's power has diminished while it continues to fetishize collective displays of self-righteousness. The only way is for an alternative *world* to come into view, that other world we have been saying is possible, and to halt the contemporary left's main activities, the avoidance behaviors of moral scolding and scattershot, reactive single-issue "realism"—unfortunately, precisely what many are advocating as "the resistance" in the Age of Trump. If the order of battle is between a nightmare world and a set of complaints, the nightmare world has already won. Moral superiority is possible but it must first be *built*; this then is the largest sense of public moral education.

An analogous story can certainly be told about those on the

political right about centrist Republicans, the religious right and libertarian types. Matters stand differently for them, though, in important ways. They have their own deep moral inconsistencies having to do with their strategic need to mesh neoliberal market-centrism with cultural nostalgia and their dangerous "paranoid style" that requires exoticized Others. For the left to compete with them on world building, though, it is neither necessary nor advisable to dispense with the self-righteousness. As deluded as it sometimes is, I am not arguing that the left should merely abandon its sense of its own moral superiority in a live-and-let-live unilateral ideological disarmament. That would be suicidal. What the egalitarian left must do is far more difficult: it must *make good on that claim*. It must move beyond critique of everyone's putatively moral shortcomings — the project of calling everyone out — and toward a process of rooting that motivating moral impulse in constructive world building. For long-term viability, the egalitarian project must ground its conviction of its own moral superiority in some kind of eudaimonistic vision capable of animating normies as well as exotics, that can structure everyday life and provide a more defined sense of common purpose. In order to accomplish this meaning makeover, however, the egalitarian left may need to leave significant parts of itself behind. It may in fact need to leave itself altogether before it can return, in a kind of hero's journey where there is no guarantee of success.

Chapter 4

Moral motivation and worldview

The crisis consists precisely in the fact that the old is dying and the new cannot be born; in this interregnum a great variety of morbid symptoms appear.
Antonio Gramsci (1935)[1]

The Nietzschean extrapolation

In order to remain egalitarian, that is, to maintain any recognizable identity at all, the actual left must distinguish itself from the identitarian visions of authenticity peddled by *both* the God-and-country (and blood and soil) alt-right as well as the "social justice" alt-left. Either path would rob the left of what makes it "left": its defining egalitarian universalism.

As a political imperative, this takes the form of a strong and abiding commitment to equality as the first principle of justice: an incomplete though situationally deployable vision of the most defensible communal life as that which is lived *inter pares* (Lat. among equals). There are nuances, of course, and continuing debates about where to draw the lines (all law and policy involves line drawing), along a host of variables including age, disability, criminality, etc., but the self-image of the left as *progressive* is premised upon an ongoing commitment toward the inclusivity of those concentric lines' circumferences. The very term "progressive" implies a directioned movement which in turn presupposes ordinates upon which the directionality could be plotted. For progressives, those ordinates are value-scaled according to a more or less utilitarian matrix of "the greatest good for the greatest number"; the more the better.[2] This is John Stuart Mill's "principle of utility," which is itself premised upon the previously described deeper equality norm of Bentham's equal consideration of interests where everyone is to count as one and only one. The historical expansion of human equality is typically conceived as a triumphal march, albeit frequently one of tears and blood, where equal consideration and its sibling civil liberties have been extended to the previously excluded: women, minorities, the disabled, gays and lesbians, transgendered individuals, children, etc.

There is a great implied faith in progress here, a secularized quasi-Hegelian version of divine providence, perhaps best epitomized in the line made famous by Martin Luther King Jr (1958) and beloved by Barack Obama and liberals everywhere: "the arc of the moral universe is long, but it bends toward justice."[3] Less loftily, in terms of everyday political sentiments, this attitude is commonly expressed by an exasperated quip in the following form: "It's [insert current year]. I can't believe we're still [insert disfavored practice]." This way of putting it implies more than just disagreement. There is a very strong sense, particularly with regard to the equality norm in at least the core areas like gender, race and sexuality (and for the farther left class), that such violations offend what should be the guaranteed historical trajectory, they go against not just the way things happen to be but the way they should be *expected* to be along their more or less normal linear trajectory. One might say that these violations represent moral backsliding; they are not just wrong, they are retrograde.

The left-liberal moral tradition suggests that, more even than whatever are the particulars of the set of "includeds" at a given time, what is really definitive of left egalitarianism is the expansive momentum *itself*, the very process of continually widening the moral circle's membership, presumably asymptotically toward an ultimate dream of total inclusion of all humanity. This dream perhaps finds its contemporary apotheosis in Facebook founder and CEO Mark Zuckerberg's (2017) ambition toward universal connectivity in order to "give people the power to build a global community that works for all of us."[4] While it may serve as a regulative ideal, this dream of including everyone is structurally unattainable because human beings are never simple generic "individuals" or, *pace* the identity-obsessed, one-dimensional "members of group X." People *per se* are never what are being connected; what are being connected are multifarious *aspects* of people. As the inherent complexities of identity affiliation show, we are necessarily internally multiplicitous. Our psyches are inevitably composite and therefore the loci of equality may flower at any time while existing loci are constantly shifting from context to context. I may discover something new about myself (a process greatly facilitated through ever-optimizing networks); what I "am" *here at this time* may shift when it becomes *there at that time* and with multivariate simultaneity (e.g., I am *this* on one online profile while I am *that* on another). The lesson of intersectionality explodes into network theory and becomes more than it ever dreamt about itself. Take the left's Big Three nodes of race, class and gender: these represent crisscrossing lines of identification and membership, and in any human community at scale they are joined by an inherently unpredictable and dynamic set of additional nodes. This ensemble then constitutes a tight weave of undecidably many connecting strands.

But the quantitative distribution of existing and relatively static moral loci is only part of the story. The other involves identity proliferation, a process that is, as earlier discussed, greatly exacerbated under conditions of liquid modernity. This means that the equality frontiers are not only external, that is, regarding the ascriptive characteristics of historically oppressed groups (e.g., African-American, female, transgender, etc.) but also they are sensitive to emergent identities as well, many of which are experienced as chosen to a greater extent. This phenomenon of the continual germination and exfoliation of identities and their accommodation is a major focus of the egalitarian left due to the abovementioned need for momentum in order to maintain that left-defining sense of constant progress. Regardless of their provenance, novel identities must be discovered and proclaimed, a process that has the ancillary function of providing an important border-defining mechanism vs. the Other Side, for example, the signaling function of the fluctuating terminologies that designate identity groups, where those using the "wrong" terms are readily identifiable as outsiders. This would include slowness or refusal to recognize emergent categories as in the transgender bathroom controversies. What becomes apparent is that the substance of the *au courant* appellations is less important than the establishment of a temporal gap by which to distinguish insiders from outsiders, the saved ones who are *keeping up* and the damned who are not. The outrage mobs are in this respect group-definitional performances.

Yet the egalitarian left is constitutionally bound to take exclusion claims seriously. Taking a step back into their conceptual origins in humanist ideology and following the broad outlines of Nietzsche's conceptual genealogy of morals, one can see that as exclusions are identified they are typically decried via an appeal to the moral substrate of a shared humanity; there is something *about* us or *in* us—all of us—that is being violated when a fellow human being is denied equal consideration.[5] At the back of this moral intuition is something still deeper: the long-hegemonic biblical tradition has instilled in us the retrofitted Platonic notion that there is something about us that is not reducible to materiality and is distinguishable even from our very bodies. There is some-*thing* to which all of our external determinations adhere and stands separable in some key yet unspecified metaphysical sense. In Platonism and later Christianity (i.e., Nietzsche's "Platonism for the 'people'"[6]) there is a recurrent motif that the body forms a "prison" for the soul. Corporality fastens the soul to the world through desire and other bodily appetites, making it heavy so that it "sinks" down into the sub-rational realm of mere things.[7] In this dualistic picture, the underlying soul-substance then becomes an essentially aporetic concept—the *"an sich"* (Ger. in itself) in Kantian lingo[7]—about which nothing coherent can be directly said but around which are grouped essential yet enigmatic notions such as immortality, free will and moral responsibility.[8]

This soul-substance is what every human being (or perhaps "rational being," as Kant would have it) has in common, deep down, beneath all of the contingencies and pollutions of the world. The soul is a great ontological leveler in this respect, as Christianity officially emphasizes in its "Sermon on the Mount" mode where "the meek" are afforded their due consideration (Matthew 5:5). In theory the Christian God does not show "favoritism" (Romans 2:11, Acts 10:34) and neither should we (James 2:1, 9). It is a short conceptual step from equality before God to equality before the law to moral and political equality (democracy) to economic equality (socialism) and perhaps to somewhere even more "equal" than even that. For Nietzsche, this development represents a slippery conceptual slope where socialism, as the apotheosis of the "equality of persons," was for him a regrettable development representing "the tyranny of the least and the dumbest" and even "a will to the denial of life."[9] He unequivocally ties socialism to Christianity, the original "slave revolt in morality": "The socialists appeal to the Christian instincts; that is their most subtle piece of shrewdness," adding also that "mankind was first taught to stammer the proposition of equality in a religious context, and only later was it made into morality: no wonder that man ended by taking it seriously, taking it practically! that is to say, politically, democratically, socialistically."[10] It is clear that Nietzsche also holds more generally that democracy itself, along with democratic philosophies such as utilitarianism, are basically Christian knock-on effects: "Democracy is Christianity made natural."[11] Despite his reputation as an anti-systematic thinker, Nietzsche thus presents a strange but mirroring obversion of Hegelian and

Whiggish triumphalism, where the development from ethereal equality before God to sublunary equality *inter homines* is viewed as moral regress rather than progress.

While Nietzsche is sometimes presented as a proto-libertarian champion of individualism, and in that sense a democratic thinker, I think this goes too far.[12] Nietzschean moral developmentalism, as outlined above, might more precisely be described as a sort of proto-Jungian process of collective *individuation* rather than as a philosophical resource for political equality, human rights or any such notions. In the Jungian tradition, individuation is a parallel process occurring at the levels of both the individual psyche and also a shared cultural process where the ego gradually differentiates itself from the hold of the unconscious. This now occurs as part of normal child development for modern persons and their maturation as integrated selves and also it can be traced as a historical development where the ego-individual emerges from the archaic and mythic "collective unconscious" of pre-history.[13] This is the kind of thing, I think, that Nietzsche is gesturing toward rather than any sort of simple ennobling of "the individual" in the current political sense.

Thus Nietzsche mostly champions heroic individuals of the creative type such as artists, musicians and poets and indeed moral innovators like, ironically, Jesus himself. The mass of humanity, "the herd," is at best worthwhile insofar as they are instrumental to the production of these few-and-far-between genius super-humans, the *Übermenschen*, these latter being the only actual true individuals. This is why Nietzsche sees socialism and democracy as so threatening: as an ascension of the "least and the dumbest," it superficially promotes individualism as political equality as a cover for a kind of mass life-hating *ressentiment* that can be deployed against the higher types as they arise among us. We will continue to murder our Socrateses and Jesuses, jail our Galileos and episodically build our bonfires of the vanities. For Nietzsche, the historical record shows that outbreaks of egalitarianism have usually functioned to depress and destroy our finest—and necessary anomalous—human exemplars. Nietzsche's real history of humanity is how these singular creative types have asserted themselves against the resentful mob. So it is too big a stretch, as some recent scholarship has, to go about "white-washing" Nietzsche and make him into an anodyne spokesperson for cheerful life affirmation and bourgeois individualism, as if he were merely counseling the masses to follow their bliss and lead more creative lives. A rather straightforwardly aesthetic elitist, Nietzsche does not care about the tasteless and dumbed-down many.[15]

It is possible, however, simply to read him against himself on the point of his *valuation* of his genealogy of morals, from the originating slave revolt in morality in Judaism and Christianity down through that lineage to democracy and socialism. In other words, one may accept Nietzsche's basic picture, as I do, yet put a positive spin on it. One need not *lament* the intellectual history of how egalitarianism unfolded. I would go one step further and argue that placing *any* external valuation on the process makes little sense—deriding it *or* welcoming it. If *ex hypothesi* one is enfolded into the very ideological process one describes, upon what normative ground would one stand in issuing such a meta-judgment? Though cagey about it, Nietzsche was quite sensitive to this problem and develops that very normative ground in various works, perhaps most notably in his compendium of homilies *Thus Spoke Zarathustra*, whose subtitle speaks to this standpoint predicament: it is to be taken as "a book for everyone and no one."[16] Despite the book's epic tirades against equality and the "secretly vengeful tarantulas" promoting it, the subtitle reflects a key ambivalence concerning to whom his writings are directed.[17] This is analogous to the earlier-mentioned problem in Marxism about how the critic (viz., Marx himself) can stand apart from the ideology-determining basis of his own society that he, after all, also inhabits along with the capitalists and everyone else. Nietzsche is more sensitive to this point than Marx. Not only is he aware of the problem of audience, he casts it as a necessary all-or-nothing gambit: either *no one* will be equipped to gain enough critical distance from it to hear his message, as we are all so shaped by the very historical processes he describes, or instead he writes to *everyone* because the same cultural cross-currents that are allowing *him* to see through what he takes to be the life-hating lies of the Judeo-Christian tradition must also make it possible for *everyone* else to see through them too.

Weirdly, despite himself, Nietzsche therefore leaves the door open, at least in principle, to a kind of moral universalism in order to preserve the coherence of his attempts to communicate his views. In a way he provides an opaque anticipation of German philosopher Jurgen Habermas's "discourse ethics," in which the very effort toward intelligibility presupposed by the act of communication implies a regard for the Other and thus constitutes a foundation from which a morally universalist ethics can be built.[18] Although politically Nietzsche is an unabashed elitist and anti-democrat, he still harbors a latent egalitarianism in that he does not rule anyone out in advance as a proper and welcome recipient of his message. He is an elitist not a snob. Conveniently overlooked by his fascist appropriators (including his own monster of a sister), he is, for example, savagely critical of anti-semitism ("I'm having all anti-semites shot"[19]), the stultifying effects of all kinds of nationalism, especially that of his own compatriots ("Wherever Germany extends, it ruins culture";[20] "The Germans themselves have no future."),[21] and wanted to have "nothing to do with a person who takes part in the dishonest race swindle."[22] One might label him a "universalist elitist" along with Plato's rigidly hierarchic vision of *The Republic*, wherein a child of any parentage is potentially a gold "guardian philosopher king," or, indeed, Thomas Jefferson (1781), who advocated for public education so that "twenty of the best geniuses will be raked from the rubbish annually."[23] Egalitarian projects can possess the sketchiest of intellectual patrimonies.

This should provide some reassurance because I want to suggest that Nietzsche's picture of how egalitarianism arose is serviceable in roughly the manner he meant it: as a kind of genealogy of ideas rather than as an anthropological thesis. I adopt it with two major alterations, though. First, I would replace Nietzsche's own negative valuation with a positive one. I see no reason to join him in *regretting* these developments. One could just as easily celebrate them. It seems to me ultimately a leap of faith either way, so I choose the leftward leap of maintaining equality as a guiding norm out of what I would candidly admit is my own baseline sentiment of fellow-feeling for downtrodden human beings rather than an impulse to venerate only the "great" ones. Second, the commitment to the equality norm must be *extrapolated* from the picture Nietzsche presents and *extended* from there beyond the arbitrary stopping points subsequently assigned by both Nietzsche's criticisms of it and also its forthright left-liberal champions. Ironically very much in the Nietzschean spirit of "overcoming," what must be overcome is what turns out to be, on examination, the equality norm's wholly unjustified anthropocentric humanism, its assumption that *human beings* are always at the center of the moral universe and that ethical care and concern—and our very conception of ourselves—must always be limned by the arbitrarily border-membrane provided by our notion of "species." The historical trajectory of equality has been what philosopher Peter Singer calls "an expanding circle of moral concern" where sentiments enlarge their circumference like widening pond ripples.[24] Reversing Nietzschean centripetalism, one may substitute an outward-pushing centrifugalism. This ongoing movement *itself* is the essence of the political left, its highlight reel consisting of particular moments of expansion having to do with human rights and the like. It is important, however, not to confuse the dance as a whole with any of its specific component dance steps. Its choreography is composed of them but it is also not identical to them.

Liquid modernity and its accompanying identity proliferation and fluidity create just the right laboratory conditions to make this kind of evolutionary moral advance possible. It would be strange to think that the millennia-old development and expansion of the equality norm would hit a brick wall and stop its outward distributive momentum at exactly this moment, at the point of "the species." If we allow ourselves to dig beneath public opinion and facile common sense intuitions, it becomes apparent that the assignment of moral concern only to other *Homo sapiens* — speciesism — is really no more justifiable than sexism or racism. Even old Kant was not subject to it, as he typically refers more widely to "rational beings" [Ger. *vernünftiges Wesen*] rather than making the unjustified assumption that the sole locus of moral theory must necessarily be *human*. Identity proliferation helps prepare the way for this Nietzschean extrapolation of the equality norm by augmenting the elasticity and flexibility in our conception of human identity, much like the stretching exercises an athlete performs prior to competition. Again it is that magician's hat and the scarves trick that has a pedagogical effect on a population that continually witnesses the spectacle of identities being birthed and gaining social traction — yesterday transgender, today that long list of +'s added to "LGBTQ" and tomorrow perhaps the "K" (kink) or some new retro-futurism or bio-engineered fusion or, more likely, any number of formations beyond current imagination.

This is why, as argued previously, one misses the essence of this phenomenon if one approaches it with the egalitarian monomania characteristic of the identitarian left. If one signs on to the Sisyphean moral game of perpetually "catching up," perhaps with the aid of the online outrage call-out machine, one will lose because the entire framework of liquid modernity *in principle* disallows any stopping point of maximal inclusion where all the egalitarian work is completed. Such a conception is a misunderstanding of the astonishing dynamism and something-out-of-nothing quality of human identity proliferation, which has no built-in stopping point—certainly neither the "racial and ethnic categories" of the US Federal Office of Management and Budget nor, from a still broader view, the equally quaint nineteenth-century relic of species as moral focal point.

The sped-up timelines only exacerbate the impossibility of playing catch-up with liquid modernity's speedup of identity proliferation over the course of an average human lifetime. Just as the elderly can appear as hopelessly morally retrograde in the eyes of successive generations due to certain attitudes and use of obsolete identity labels (e.g., "colored," "broad," "Chinaman," "tranny," "illegal," etc.), the increasing rapidity of identity proliferation will shrink the time-windows for catching up such that almost everyone will start appearing to everyone else as obsolete and morally "out of it," 30-year-olds will seem to 20-year-olds as 80-year-olds now seem to the 30-year-olds—analogous to how a current phone app like Snapchat seems almost conceptually out of reach to anyone over 40. This shrinking of temporal horizons, an obvious structural aspect of an increasingly on-demand economy of instant consumer gratification (2-hour grocery delivery via Amazon!), is an inherent aspect of identity proliferation; the proliferation is temporal as well as spatial. It is thus impossible to get caught up; everyone is inevitably "behind the times" and even the most egalitarian and inclusive saint imaginable will reach her limits with regard to assimilating the next new thing. Like commodities, we will all experience *moral* obsolescence in this way, a gestalt shift constituting a fundamental reversal from most of human history where the default tendency was to look to elders for wisdom and guidance. The sixties meme of "never trusting anyone over 30" thus becomes reduced to a terminal absurdity where everyone withholds full regard for everyone else merely because we are all temporal and mortal beings. None of us will ever "get with it."

Instead of trying to find a way out of this predicament or aiming to slow down the moral carousel to a comfortable speed—which is impossible because the very forces causing the situation are only growing as per the liquid modernity thesis—I would argue that something like an "accelerationist" mindset on this point is the only conceivable desirable option. The only way out is going to be *through*. This involves recognizing that the identity proliferation volcano is going to continue to erupt both more voluminously and more frequently. The moltenness of the identity lava is not going to be stopped and it is futile to try to do so. The project of egalitarian accommodation of the relentlessly emerging identities will sooner or later break even the most devoted and over-socialized inclusivist identitarian. I think that the only way to bound this process and to make it psychologically manageable is to ease the collective introspection and begin to look *outside* of ourselves in order to regain the egalitarian momentum. As I have been arguing, it is the *motion itself* of the equality extension that is necessary for the left to sustain itself, rather like how some sharks need to swim perpetually forward in order to breathe. The Nietzschean extrapolation, then, is to de-emphasize the application of egalitarian norms to identities as they arise *within* human beings—as was done on a historical scale previously in the soul-initiated Christianity-democracy transition—but rather to *extend* their application to realms *outside* of humanity. Though he himself did not imagine it, turning away from anthropocentrism would be a kind of moral "overcoming" in the Nietzschean sense.

As mentioned earlier, Kant had the jump on everyone with his deliberate designation of "rational beings" rather than "human beings" as the agents and objects of morality. This was a strangely prescient move on his part and it should give courage to even the most ardent humanist to consider the merits of detaching "morality" from "humanity." Kant's terminological choice is consistent with his emphasis on moral autonomy as the acceptance of self-given laws and allows his perspective much more easily to extend equal consideration to, for example, great apes along the lines of primatologist Frans de Waal's compelling analyses of ape moralities, and could support such initiatives as Spain's push for great ape personhood.[25] Within the Kantian paradigm, one might include actual creatures such as dolphins, octopuses and ravens along with possible creatures like space aliens and certain AI robots, all of whom could easily qualify as rational beings in any meaningful sense. But there is really no reason to stop there or to insist specifically on "rationality," a term so notoriously difficult to define that it is philosophically nearly useless. If the goal is an extrapolative normative push beyond humanity and "rational beings," the contemporary animal rights movement provides an example of an extension of moral concern beyond our own species. Such perspectives erode the Cartesian dualism of mind and matter, where nonhuman animals and nature generally are consigned to the latter category as mere *res extensa*, a vast system of inanimate (i.e., literally soulless) mechanisms that as such lie outside moral concern.

But again: why stop there? Why stop at nonhuman animals or even "sentient beings," to use the wider Buddhist phraseology? Or, as environmentalist philosopher Arne Naess further suggests, that "care flows naturally if the self is widened and deepened so that protection of free nature is felt and conceived of as protection of our very selves"?[26] But does extending concentrically further outward land us in some ridiculous position that is inimical to human life, given that we are no exception to the biological reality that life must devour life in order to survive? Even the most scrupulous vegan leaves a vast trail of death and destruction in her wake, modern agriculture being inherently violent toward land and soil, a holocaust for countless ancient ecosystems.[27] And there is a further caveat about latent misanthropy apparent in the notorious claim by one prominent "deep ecologist" praising Ethiopian famine as population control.[28]

How might the needed extrapolative egalitarian momentum be maintained along with the equally necessary need for limits? Is it possible to maintain *some* limning non-moral realm on the outside so that the field of moral concern does not just become washed out indiscriminately into everything? In the infinite run perhaps such totalism is possible, and maybe there are mystics who have enjoyed ineffable experiences of it, but at present saying one must care about *everything* means that one need care for *nothing*; analogous to what Edmund Husserl's phenomenology held about the inherently intentional nature of all consciousness, its object-directedness: all caring is caring *for* something, there is no "pure" caring *per se* separable from its object.[29] One never just simply *cares* (or hears or sees or whatever); one always *cares-about* X, Y or Z (or at least what one *takes* to be X, Y or Z). And if care qua moral solicitude is intentional and must always make reference to some-X—in other words must have a determinate character—then this means that if it is to preserve any definition then that to which it does *not* extend must be co-present as well. For there to be a moral circle at all there has to be an *outside* to it as well. Apart from some mystical vision of oneness outside logic and morality—a possibility that would be foolish to discount—one cannot maintain an inside without an outside.

While maintaining their ethical momentum requires egalitarian concerns to be extrapolated outward into a larger circumference, those concerns still must be fixed onto *something* in particular. The *telos* simply cannot be "care and love for *everything*," as this would depart from the realm of conceptual understanding upon which ethical commitments depend. There are of course motivating sentiments involved but in the phenomenological manner ethical actions always require intentional objects. If there were no object of thought involved then the action would be purely reflexive or random and could be described as "ethical" only by external accident, as in a case where salutary consequences happened blindly to ensue. Ascribing moral valence in such situations would be a category mistake, akin to morally praising or blaming a thunderstorm.

The world's great wisdom traditions all contain mystical visions of oneness that trail off into ineffability and provide a perspective from which distinctions and divisions ultimately do not matter. These are outside my present scope and competence. It does seem certain, however, that in the sublunary articulable realm of ethics and politics *something* determinate is required upon which to fix normative commitments and aspirations. This something requires a great deal of attention, for adjusting moral perspective is not a disembodied armchair intellectual exercise whose conclusions are later applied to a static world. It is a more holistic matter where the world within which an individual's perceptions are to be meaningful must be conceived as dynamic and existing before, during and after those perceptions. In other words, it must somehow be built beforehand and then maintained, socially co-created as an ongoing intersubjective educational endeavor in the grandest sense. The greatest educational need of the useless people we are becoming is *to live in a world that matters*. We need to see ourselves *as* something that is not just a repackaging of our individual selves. Human beings must live larger than that.

World building options

Basic rationality, whose most concise definition involves the adequation means to ends, requires a critic of some state of affairs ultimately to *come from somewhere*. This is why the online outrage mob so easily turns into the apotheosis of bad faith: its anonymity and pure negativity allow it not only to hide personal identity but also relieve it from the responsibility to disclose the perspective from which the outrage issues. In place of explicitness and candor one gets a sub-rosa emotive egalitarianism that allegedly should be "obvious" to any right thinking person—the obvious corollary being that if it is not obvious to *you* (likely due to privilege or prejudice) then maybe #youtoo are part of the problem and you need to be called out and shamed. If this form of social critique becomes systematic it becomes formally irrational in that, absent any implied social consensus on the normative basis of the call-outs, the means being used (i.e., the public denunciations and shamings) are not being adequated to any particular substantive ends; the call-outs have no grounding other than the transient emotions of the shamers. It is not that one needs to provide a fully articulated worldview in every tweet, but rather that *at some point* down the line something approximating a premising worldview should be articulated (or at least pointed toward) to show from precisely what standpoint the outraged one derives the outrage. This is not an onerous requirement. In fact it is arguably the default historical mode of social and political critique, for the customary etiquette of denunciation is to do so *in the name of one's worldview.*

Since the online world is so quick to condemn others as fascists and Nazis, here is an example of a proper Nazi denunciation in the traditional style. It is from the 1939 British anti-Nazi propaganda film *Pastor Hall*, where the eponymous protagonist confronts a leading Nazi official with these words:

> you and I stand face-to-face unmasked in the sight of God, the words I speak will belong to another not to me but they shall be spoken. . .*I denounce you in the name of God* and with you I denounce the rulers of this country, their whole system, this vile growth bred in darkness and in hate, which tortures bodies while it murders souls, decrees men should be kicked and beaten like the best of the field. . .flogged to death while they whisper the words of God. . .I denounce this Hitler,

architect of evil, creator of human misery [emphasis added].[30]

It runs longer than the requisite 150 Twitter characters, but this is how one does it old school: the intensity of the denunciation is matched and palpably authorized by the clear evocation of a worldview. Neither the denunciation's recipient nor any observer could have any doubt about precisely where in the normative universe Pastor Toll is coming from. This is no mere call-out.

Liquid modernity and its proliferation of identities have made it more of a challenge to "come from" somewhere that is broadly intelligible, though. As per my foregoing hypothesis, as the egalitarianism extends itself through the identity explosion, it is less and less possible to base social critique on any shared eudaimonistic vision, whether that be medieval Christendom, communist internationalism, Islamic Wahhabism, Mussolinist fascism, machine-human singularity, Native American Waashat Dreamer Faith, ancient Greek Orphism or mega-church prosperity gospel. All the while, the enabling condition of this fragmentation, the reigning ideology of neoliberal market fundamentalism, is no help due to its formal nihilism that contains no vision of the good life at all; as a poor stand-in, neoliberalism has only a very thin set of board-room managers' assumptions about human nature (viz., human beings are rational self-interest maximizers), that it has transposed from highly artificial neoclassical economics models.

In this sense, neoliberalism is not really an ideological antagonist in the first place, as it takes two to tango; its thin set of underlying assumptions obviates any actual clash of ideologies in the manner of World War II or the Cold War. In order to separate from this mindset and to operate once again at a basic threshold of rationality, it is necessary to engage in that most "retro" of human activities: gesturing at the stars, looking upward from ourselves and trying to narrate a conception of the "whys" of our existence that can defensibly premise moral choice. Again, if rationality is conscious adequation of means to ends, and we want something more than to be led around by consumer preferences or kneejerk emotivism, it is incumbent upon us to examine matters from the perspective Kant called "the kingdom of ends," a realm of ideals that will provide the major normative premise for whatever are our best-laid plans.[31] If such high-mindedness holds no allure, greater contemplation of one's worldview starting points should at least provide more bite to denunciations of one's ideological opponents—because then one will really mean it. Ironically, it is strategically much more effective actually to stand for something.

If Bauman is right that liquid modernity can be only an interregnum because human beings, as narrative creatures, can only stand formlessness and directionlessness for so long, what is all-important for the future are the worldview molds into which the liquid identities eventually be poured.[32] This is the real ideological struggle—what Detroit social justice facilitator adrienne marie brown calls an "imagination battle"—not the online outrage *du jour* or the *ad hoc* policy debate of the moment (as of this writing it is gun control again, soon it will be another in the nonstop parade of media ephemera). So what then might be the options? What are the worldviews that might conceivably be built or rebuilt?

By "rebuilding" I mean somehow reconstituting a moral universe with reasonably universal (or at least hegemonic) adherence, one that simultaneously "goes high" yet also "goes low." It is relatively easy to hit one or the other of these inflections. A durable worldview, however, needs both simultaneously. The "going high" part has to do with the providing of at least provisionally satisfactory answers to the Kantian troika of perennial ultimate questions of God (or Spinozan God-or-Nature or some other marker of cosmic ultimacy), freedom and immortality. One may find synonymous stand-ins for these three

149

but the point is to provide a belief system, a *Weltanschauung,* that says something to our ultimate purposes as individuals and as human or "species beings" (in the Marxist tradition) and/or as *some* kind of entities who have an identifiable place and purpose in the cosmos. Derivative from this ontology is a philosophical anthropology consisting of some set of assumptions about what we are as human beings, what constitutes the good life (*eudaimonia*) and, further, what we can and *ought* to do in light of those descriptions. And finally, in light of the ineradicable reality of sublunary impermanence, *what, really, is the point*? What happens after we die? Why bother with striving at all? Some quarters of university-based philosophy still take up questions like this but they have become largely unfashionable— even sometimes embarrassing—in the busy world of academic careerist everydayness. Even at their best, the philosophers who consider such questions are not out to offer integrated wholes; the academic division of labor consigns them to offering highly specialized answers for highly specialized audiences. They are high-end niche designers selling their designs only to other high-end niche designers. Perhaps the fashion trends may trickle down to the public, perhaps not.

Within the history of religion there are institutionalized analogs to the going high strategy where high-end theologians and their esoterica have commonly been integrated into larger wholes—one of the most important functions of institutionalized religion. Catholicism, Orthodox Judaism and both Sunni and Shi'a Islam have long traditions of authorized interpretations where impressive spaces are carefully maintained for the most high-end and abstruse theological debates. These differ from the secular academicians' reveries, however, in that despite their customary delicateness, clerical elites are more holistically connected to a larger community of believers, they have their place within an encompassing vision of the whole, their *Ummah,* as it were.

Outside the ivory precincts, the "going low" requirements for an operative belief system crucially involve creating a patterned rhythm for the quotidian, rituals, seasonal customs and life cycles. Birth, death, coupling, sexuality and family-formation, crime, punishment and expiation, holidays, festivals and communal entertainments. And of course formal education and all modes of social reproduction: the traditional assumption is that these things need to be tethered to a comprehensive conception of the good lest they degenerate. As human beings we are not satisfied with stand-alone QED philosophical or theological conclusions; *we need patterns actually to live by*. We need to make sense not just of Kant's "starry heaven above me" and "moral law within me" but also to *regularize* the events and cycles of our lives within some renewable and intergenerational framework of ritual and meaning. Continually in the throes of self-narration, people have to be able to make sense of their lives, not just in that high cognitive sense of Big Answers to Big Questions but in the low sense of durable and repeatable enframings of tonight, tomorrow and next year—including the sacralization of major life events: births, deaths, rite of passage, marriages, harvests, seasons, commemorations, etc. "What festivals of atonement, what sacred games shall we have to invent?" declaims Nietzsche's madman.[33] Wisdom counsels that the small things must be given their due. In their ensemble, their repetitions and patterns give rise to feelings of worthwhileness and meaning for an individual and are necessary conditions for any form of communal life.

There are many perspectives from which to appreciate these quotidian necessities. There is a scene in Woody Allen's *Manhattan* when, lying on the psychiatrist's couch, he lists the things that make life worthwhile: "Groucho Marx, Willie Mays, the second movement of Mozart's *Jupiter* symphony, Louis Armstrong's recording of *Potatohead Blues*, *Sentimental Education* by Flaubert, Marlon Brando, Frank Sinatra, those incredible apples and pears by Cezanne, the crabs at Sam Wo's, [his girlfriend] Tracy's face..."[34] I suspect that most people resonate with Allen's character's general sentiment; meaning is not necessarily found in the grand conceptual realization or the cathartic spiritual breakthrough but in life's little things whose aggregation adds up to more than their sum. Ultimately there is nothing to be denigrated in locating meaning via the going low strategy (like Allen's character), which involves cherishing life through its *moments* as they have been encountered along the way. It is a phenomenological vision of life as it is lived from the *inside* rather than the altitude of a bird's eye view.

As they are disproportionately intellectual types prone to the high-end way of looking at things, political activists should be mindful of human life as actual people actually live it rather than as a play of abstractions. As a philosopher I know I am prone to this error as well. Too often we are presented with a love for humanity merely in the abstract—sometimes this is combined with contempt for actual people—and designs for the beautiful castles of sand they are accommodatingly to inhabit. From their commanding heights, statist planners can forget that, while the larger and farther vision is indeed necessary, much of life's worthwhileness lies in the interstices of the larger structures, in the quieter and more intimate spaces usually left untouched by political schemers. Allen's list of person-scale worthwhiles is a reminder of this.

Still, the consoling lexical hedonism of the cosmopolitan aesthete is neither materially possible nor psychologically desirable for most of us by itself. Most people need their life's puzzle pieces also to fit into something more or less coherent; as narrative creatures we have a need to see ourselves as fitting into an overall pattern that extends beyond our narrow selves. Since most of us are not highly original thinkers—or more charitably put we are social creatures in *all* of our aspects including our mentalities—we are also more often than not content for our worldview to be *received* from inherited traditions and those alongside us. This is overwhelmingly and perhaps, in the end, universally the human norm. An imagined libertarian dream space of individualists who have autonomously built up their own singular worldviews is just not bloody likely. Really, it is impossible. Instead, we necessarily integrate ourselves hermeneutically into pre-existing languages, traditions, ways of life, customs and habits that we ourselves did not personally fabricate. Without these integrative contexts, the individual's experience of life—if it would be imaginable at all—would consist of a meaningless mash-up of Allenian ad hoc experiences and scattershot impulses.

So what could fill this tall integrative order? What could possibly satisfy both the higher and lower needs? I see four broad possibilities for western MEDCs that are sufficiently "Big Questions"—focused yet also conceivably operable at the qualia level of individuals' first-person experience of their own lives. These worldview options also enjoy a degree of cultural rootedness and therefore potential scalable resonance, which is a necessary condition for them to be considered realistic options. History shows that nothing invented out of whole cloth—and nothing *merely* philosophical, no matter how argumentatively compelling—is going to have much chance of catching on. The most resonant worldviews are built of materials that are largely prior to political ideology and, like the major world religions, in principle could be taken in whatever political direction. If one were to sign onto one of these worldview options, there is no guarantee that it will end up in accord with any particular set of predetermined political outcomes. The die will have been cast.

(1) *Traditional Biblical Religion.* In the US this means Christianity (with Judaism, Islam and fusions with other traditional world religions—particularly Western Buddhist and Hindu imports— inhabiting constitutionally-protected niches of tolerance). This

general approach of course far predates capitalism and will outlive it as well. Its major liability is that its central tenets are obviously false if taken literally; to the modern skeptical mind the main components of traditional Christian theology (e.g., original sin, virgin birth, resurrection, heaven and hell, angels, miracles, Satan, etc.) are childish if not ridiculous and simply cannot be taken seriously—more like the products of some kind of hallucinatory mental illness than understandable as descriptive of reality. It is more intellectually palatable therefore to understand the theological contents of these traditions metaphorically. Yet adopting this interpretive mindset raises serious questions about why these traditions deserve to be privileged in the first place (and by extension the hierarchies of interpretive authority—"the Church"—that inevitably ensue). Why all the smoke and mirrors? Why not just take a more honest straightforward approach?

(2) *Identity-based authoritarianism.* This pattern of worldviews is premised on a "thick" group identification (and/or nationalism) that tends to be ultimately based—explicitly or implicitly—on a correspondingly deep essentialism (biological or historical). This is an often bellicose and structurally unstable form of consciousness because it always requires an Other against which to define itself. This tendency can become extremely dangerous as allegedly life-and-death struggles are a feature not a bug of large-scale identitarianism. But qua nationalism it is also foundational to modernity, e.g., the formation of the nation-state in the early modern period, and has unquestionably proven able to provide deep satisfaction for some people. On the political right it is a "blood and soil" proto-fascist mythos concerning "a people" realizing its allegedly essential nature and on the left it is an identitarian "social justice warrior" tendency vis-à-vis historically oppressed groups whose aspirations toward their own nationalist "self-determination" have been wickedly

thwarted. Both represent an interestingly bidirectional identity movement: at once outward and wider beyond family, blood-feud and tight localism, yet also at the same time a rejection of extreme humanist universalism (e.g., nationalism and racism represent relatively wide identity-referents; there are a lot more "Aryans," "Italians," "Palestinians" or "Latinos" than there are in one's band or village). Yet both nominally conservative and progressive identity agendas are ultimately based on a politics of authenticity and as such are inherently hierarchical (some individuals are always established as "more authentic" than others). As identity-grounding worldviews, these perspectives typically arise under conditions of oppression and/or threat or *perceived* oppression and/or threat. The identitarian-nationalist mentality runs deep and is probably impossible wholly to dislodge from collective consciousness.

(3) *Techno-aesthetic futurism.* This category includes a great variety of both aesthetic and technological (including bio-tech) futurist visions with explicitly comprehensive ambitions that look beyond their originating spheres of art or technology. These are visions that seek to change the world as a whole rather than just innovate within the narrow range of their own disciplinary practices. This aspirational component must be stressed in order to distinguish this set of worldviews from merely intramural "schools of thought" or from fads and fashions that may periodically sweep through their larger host societies. Though well-represented in popular culture, the various futurisms tend to be highly intellectual undertakings that attract visionary eccentrics in the arts and sciences. In terms of numbers this is the tiniest of groups, though their cultural influence can be outsized due to their innovativeness and the aura of excitement they often generate.

(4) *Archaic religion (pre-Biblical).* i.e., some nature religion

or ecosystem-oriented neo- version of animism, paganism, Druidism, Wiccan, Gaia worship, along with various worldviews typical of modernity-resistant indigenous peoples worldwide. Scholarly sources for these belief systems may be found in anthropology and miscellaneous other disciplines, such as in the vivid tableaus presented by Jungian psychological comparativists like Mircea Eliade and Erich Neumann, who have examined cross-cultural mythic archetypes. Philosophical antecedents of this general approach include certain strands of Stoicism, where contemplation of nature is regarded as an important pathway to *ataraxia* (tranquility) (e.g., Lucretius's classic poem *On Nature*) and Baruch Spinoza, who argues for a radically synthetic conception of "God-or-Nature" (*Deus sive Natura*).[35] There are also non-archaic visions and practices that might plausibly be placed in this category in the sense that they are eco-centric and tend to exemplify what religion scholar Bron Taylor calls a "dark green religion" that "considers all species to be intrinsically valuable" and stress: 1) the kinship of all life, 2) human humility and non-anthropocentrism and 3) the interconnectedness among branches of science (e.g., ecology and physics).[36] Further within this category is a range of approaches Taylor labels "Gaian Earth Religion," where the biosphere is conceived as a self-regulating hyper-organism that is "alive or conscious, or at least by metaphor and analogy to resemble organisms with their many interdependent parts."[37] Literally more down to earth, this general outlook is often associated with certain consonant horticultural practices such as organic farming, permaculture and edible forest "food gardens."[38] There are also radical environmental and animal rights activist groups whose belief systems flow from identical or kindred sources. This category will be the subject of Chapter 5.

Traditional biblical religion

American right-wing orthodoxy proceeds as if the homilies of desert dwellers thousands of years ago present a perfectly coherent eudaimonistic blueprint for today. As such this worldview faces severe challenges processing a complex heterogeneous world, most of which cannot share its irrational and arbitrary theological premises. Overall it depends on an enforced magical thinking that seems not built to withstand the corrosive skepticism of the modern mind, the fated incredulity toward metanarratives and total explanations that are supposed to be taken "on faith." This includes what historian Bart Ehrman identifies as the clear cause of the "triumph" of early Christianity over competitor religions: its willingness to rely on supernatural "signs and wonders," i.e., miracles, as "evidences" (sic) of its unique superiority.[39] Christianity does not just *include* certain irrational aspects, as perhaps does any popular belief system, it is *based* on them, its occult elements probably constituting the core of its mass appeal. (This analysis is consistent with Dewey's critique of biblical supernaturalism as outlined in Chapter 2.) Systematic coherence thus takes a back seat to whatever emotive forces happen to well up from the cultural fissures of the day.

Typically, then, with regard to politicized Christianity, what one finds is merely religious-themed posturing with little internal coherence, as when fundamentalists adhere to pre-existing prejudices to cherry-pick scriptural support for a cultural agenda—based on nothing but their unexamined feelings of disgust—in areas like the strange Leviticus denunciation of homosexuality, while ignoring other inconvenient (and patently ridiculous) but equally scriptural admonitions such as occurs just *one section later* in Leviticus forbidding the wearing of cloth woven from two kinds of material.[40] Since no sane modern person can live by *all* of these precepts, the caprice of the ensuing selectiveness demonstrates that these religious movements are really *cultural* movements for which religion functions as a mere pretext, however deep and unacknowledged the culturalist motivations may be. Ironically, what is taken on faith is not so much the religion as it is the capriciously-received cultural prejudices, impulses that are far more potent than the biblical doctrine everyone allegedly cares so much about. In terms of contemporary politics, the opportunistic Trumpism of so many evangelicals only bolsters this conclusion. However one characterizes the Trump phenomenon—and even if one thinks that it might signal a kind of hidden rationality, perhaps something along the lines of a noble savage populist rejection of elite business-as-usual—the idea that Trump himself somehow represents anything close to "biblical values" is patently ridiculous.[41]

Despite the modern American religious right's opportunism, it is important to remember that all of the world's major religions far predate capitalism. They even predate capitalism's predecessor, feudalism. It would be foolish therefore for the would-be revolutionary to assume that religious forms of understanding will simply vanish along with the hegemons after an imagined overthrow of capitalism, as Marx seems to imply with his famous religion as "opiate of the masses" remark. On the contrary, it is to be expected therefore that, whatever their contemporary corruptions, there is a high level of protean resilience lying within these intricately layered traditions. Although urban progressives tend for cultural reasons to view them as retrograde they cannot be written off so easily. Generations of official atheism under the Soviets reduced belief but could not eradicate it. And modernity only seems to fuel various Islamisms in the Middle East and elsewhere.

There are forms of Christianity right in the belly of the neoliberal beast that seem already to point once again toward something beyond capitalism's consumer ethos. An unlikely but stunning example is a strain of evangelical "radical Christianity" now gaining popularity that quite vociferously attacks the greed and materialism of "the American Dream." Alabama Pastor David Platt preaches that the greed and materialism of what he identifies as "the American Dream" are fundamentally at odds with the spiritual teachings of the Gospel.[42] Platt summarizes his standpoint: "Real success is found in radical sacrifice. Ultimate satisfaction is not found in making much of ourselves but in making much of God. The purpose of our lives transcends the country and culture in which we live. Meaning is found in community, not individualism; joy is found in generosity, not materialism; and truth is found in Christ, not universalism."[43] Due to their own cultural prejudices, an evangelical church in a mostly white area of suburban Birmingham, Alabama is the last place many would expect to find a radically anti-consumerist and anti-American Dream message. But there it is. Clearly, Christianity has many latent possibilities, some of them quite at odds with what has come to be its political reputation.

Religion is after all a cultural universal. It would be exceedingly strange to think, along with Marx, that a change in the mode of production would bring about humanity's release from it, as one might pass through opium addiction withdrawal without at least a little methadone, or *something* to ease the existential craving. The closest we have come are perhaps secular religions such as those, at least for a time, under the various communisms and fascisms. The Soviet system, for example, had both high and low. There was of course a mechanism authorizing a canon of high theory, centered on the writings of Marx and Lenin and others. (The major religions all have a theological, usually priestly caste-dominated, superstructure in which the esoteric questions are addressed by a hermeneutical tradition of authorized interpretations.) There was also an everyday popular culture of sanctioned holidays, entertainments, public art and architecture, education and so on. In a way there was also something akin to an answer to questions of ultimate purpose that traditional religion excels at providing (e.g., pleasing or imitating or returning to God): a teleological conception of history where human beings are perfecting themselves through communism, it allegedly being the fullest realization of our "species being," as Marx put it. It also included the requisite *paideia*, a conception of the ideal human being, whose earthly existence and reproduction constitutes an inherent good. Channeling our energies toward the fabrication of this Socialist Man and the blood and tears shed for him was to be every bit as inspirational as that of *Imitatio Christi*. Just like the religious traditions it was supposed to have replaced, old style communism was designed to function as a comprehensive conception of the good, regulating daily life, ordering the virtues and providing answers to questions of ultimate purpose. Its relative merits as such are beyond my scope, but its ideological ambitions in this respect are clear. It was meant not merely as an economics or a politics but as an encompassing worldview that could provide holistic satisfaction and long-run direction for human beings.

It should also be noted that this high velocity of worldview expansiveness was also present in the leading fascist ideologies, at least the ones that tended to erect their own mythos, such as the Nazi obsession with Teutonism and the Mussolinist evocation of ancient Rome, both of which premised an entire system of living on a design to align all societal institutions (*Gleichschaltung*) and dissolve individuals into a visionary whole. (In this one might then differentiate Nazism from, say, Francoist Spain, which was much more ideologically intertwined with elements of the Catholic Church.) It is not that Christianity, communism and fascism are morally equivalent (if one wants to create scorecards) but that their imaginative scope is analogous; they each create ordering frameworks of value by which individuals can set their ideological clocks. Similarly, they have each tended to be highly elaborated *in situ*, that is, they create institutional hierarchies in order to develop and interpret themselves, fascist propaganda offices, for example. This is so because the core beliefs are to the modern skeptical mind so obviously farcical: neo-Teutonic and neo-Imperial Roman grandiosity, the supernatural elements in traditional Christianity, etc. These are not ideological systems that are self-evident or able to withstand the critical scrutiny of any reasonably autonomous mind. A lot of work must be put into advertising, disseminating and maintaining them. The rebellious energy of the good Pastor Platt is likely to be coercively contained within such authoritative interpretive structures that are, in fact, designed specifically to ensnare such overly zealous moral innovators. But one never knows. Jesus himself enjoyed some posthumous success in this area, as he seems to have gone quite viral.

Today's aspiring Christian revivalist, however, faces an audience whose minds have been forged by an Enlightenment empiricism that might really *like* to believe in miracles and the supernatural but cannot quite really do so and less still live accordingly. This is not to say that there are not still fervent religious believers here and there, of course; old patterns of thought can die hard. We admittedly adhere to *successor* myths, for example, a myth of progress, perhaps, but on the whole the traditional Judeo-Christian stories just can no longer be taken at face value and the vast interpretive machinery needed to square them up only makes things worse. Absent a new Dark Age, during which we all somehow forget everything from the last 5 or 6 centuries, it is hard to see how traditional biblical religion once again takes mass hold of our collective mentality.

Perhaps along with some fusion with Buddhism and/or Hinduism, both of which have had a strong Western presence since the 1960s, the three main biblical religions have shown themselves over the millennia to be remarkably resilient and adaptable. As the growing worldwide popularity of Islam demonstrates, particularly in the developing world and among immigrant communities in the MEDCs, it would be foolish to rule them out with too much confidence. Though it is perhaps imaginable, as in French novelist Michel Houellebecq's febrile but vivid (and paranoid) near-future scenario of a Muslim electoral takeover of France in 2022,[44] and despite the ongoing theocratic reveries of the powerful American religious right, it is still difficult to imagine a *scalable* revival of old school biblical religiosity among the relatively educated and pop culture-saturated denizens of the MEDCs. One never knows though. As is a major theme of this book, the widespread nihilism wrought by neoliberal capitalism provides fertile soil for all sorts of potential ideological flora, exotic and otherwise.

Identitarianism

By "identitarianism" I mean a worldview anchored in the conception that what ultimately matters is *membership in a tribe, nation or people*; and one's identity and sense of self derives first and foremost from that membership. It is important to recognize that an affiliative first-person plural mindset did not arise with contemporary politics. It is of ancient provenance and is the default setting of human subjectivity for most of our existence. Unsurprisingly, this category is even more diverse than the archaic nature-oriented group and there are plenty of intersections with religion and other categories (e.g., Serbian-Orthodox v. Croatian-Catholic v. Bosnian/Kosovar/Albanian Muslim). The common thread, though, is that the primary identification is to some concrete group of actual living people, a more or less *völkische* conception, traditionally linguistically-based (as German philosopher Johann Gottfried von Herder was among the first to emphasize), as against, say, an imperial subjecthood, theological allegiances that might be transcultural or transnational (e.g., Christendom or the Muslim Caliphate) and, certainly, the de-affiliative individualism that arises in the modern period characterized by historical liberalism.[45]

When it comes to more recent identitarian conceptions, a characteristic ideological move is the positing of an Ideal Type (implicitly or explicitly) along with various rites of purification along with a vague telos involving some kind of putatively harmonious ethno-state. The metastasizing vision of the "Thousand-Year Reich," with *lebensraum* for well-qualified Teutonic types, is an extreme version of this tendency. This kind of associational eudaimonia admits of an almost infinite variety, however. The stylized "Kingdom of Wakanda," as depicted in the recent blockbuster film *Black Panther*, "*Star Wars* for black people," according to writer Ta-Nehisi Coates, who revived the old Marvel comic book series, being a case in point.[46] In this respect it is interesting to note how, to the surprise of naïve progressives, the identitarian alt-right by and large embraced Wakanda as an African version of their own ethnostate aspirations.[47] In the film, Wakanda ultimately evolves into a more universalist vision as part of its Hollywood ending, but its alt-right enthusiasts predictably ignore that part. Because of its perennial ability to attach itself to the issues of the moment — there are always actual or latent group aspirations in any heterogeneous polity — identitarianism takes on the appearance of novelty to those with short time horizons. But this approach is at least as old as Jerusalem and Athens: the covenantal theology of the Old Testament that culminates in the formation of Judea in biblical times and the pan-Hellenism of the ancient Greeks, as embodied in the Olympic games and the Persian wars.

All that is required is a self-conscious identification (ascriptive and/or voluntary) with some identifiable group of whatever kind. The group can take just about any form imaginable — races, ethnicities, cults and tribes, as well as gender, sexual orientation and all manner of others. As a politics, the identifications tend to be sharpest around aggrieved groups who wish to exert demands on the mainstream. "Identity" is not a precise concept, though, and borders and definitions tend to be fuzzy. This is perhaps why this category's core concept of membership is typically quite fraught, to say the least, from the dismally elaborate racial eugenics of the Nazi Nuremberg laws to categorizations of who is "really" black, white, Latino, American, Jewish, gay, Serbian or whatever. The membership criteria may be conceived as biological, "socially constructed," voluntary, or some admixture of these three. I think it is safe to say, however, that historically, these kinds of identities are usually ascriptive.

Not always, though. Ideological coherence is hard to come by here, as different political orientations mix and match quite promiscuously with different ontologies of identity. Though proponents sometimes want to portray them as anciently rooted and stable, their morphology is inherently unstable and dynamic. Biological or other essentialized designations can seem oppressive, even murderous (e.g., the Nuremberg laws) or they can be perceived as liberating, as in the case of sexual orientation where homosexuality is adamantly conceived by insiders as not at all a choice. Confusingly, what might be taken to be the politically progressive ontology for sexual orientation is alterable even within that broad category. For example, whereas it is clearly an intensely-held belief that homo- and heterosexuality are not voluntaristic, and also that gender identity is not really a choice (for both trans- and cis-gendered individuals), it is increasingly recognized that "gender fluidity" is to be respected and that such individuals may identify as masculine at one time and then "switch" to feminine at another, where this switching may occur at any time—a phenomenon that surely is operationally indistinguishable from *choosing* to identify with one gender and then another—or none at all in the case of those identifying as "asexual." Furthermore, as theorists of intersectionality have emphasized, the ascriptive nature of many identities, and thus one's essentially involuntary participation in them, are, *qua ascriptive*, factical social givens that one does not choose in any meaningful sense. More, there could be a simultaneity where one can be both at once. This phenomenon seems relatively common: African-Americans and Jewish-Americans are both categorized as such by external—and sometimes persecuting—*others*, including *de jure* discrimination under past Jim Crow racial segregation laws or European laws restricting Jews. Members of oppressed groups may also at the same time voluntarily *embrace* those same ascriptive identities as in the case of, respectively, Afrocentrism/Black Power or early Zionism.

So it is impossible to generalize about the ontological status of these designations. What may be said is that for any given identity designation, there is always *some* conception of its ontological status and provenance among insiders, though there may perhaps be persistent internal disagreements, e.g., the now-material question of "Who is a Jew?" in the context of Jews making *aliyah* to Israel (a legal right under the Israeli Law of Return). Is it a Jewish mother or father? Is a grandparent enough? A Bar or Bat Mitzvah? Growing up "culturally Jewish"? And so on.

The identity politics of both the right and the left is inherently unstable and contextual. A "blood and soil" mythos could be regarded as reactionary and horrific (Nazis) or as "liberatory" (Palestinians or Native Americans) depending on the context of perceived oppressive circumstances. On the surface, left identitarianism typically presents itself initially as an egalitarian demand for inclusion, as in American racial desegregation or the granting of rights of access to disabled or transgendered individuals. Because identity provides group cohesiveness, though, the internal makeup of such groups typically develops along hierarchical lines where, in this respect like Plato's *Republic*, proximity to an ideal—in this case a cultural ideal of racial/ethnic/tribal/etc.—is the touchstone of legitimacy and authority for individual group members. The closer one is to the touchstone the more authentic one is. Consequently, an ethno-nationalist leader of whatever type would generally be expected to have a solid pedigree of some sort in order to establish his or her authenticity in inhabiting the ideal. Problems of legitimacy will characteristically arise as a function of proximity to some touchstone in this sense. This is why, left unmitigated, identitarianism in whatever packaging always crystallizes into an authoritarian hierarchy. Analogous to biblicalism's reliance on canonicity, its underlying moral geometry of authenticity-proximity guarantees it.

A telling recent example is that of Rachel Dolezal, President of her local NAACP chapter in Spokane, WA, who was found not "actually" to be black and had been "passing" through a variety of means including cosmetics, hair styling and a fluency with the relevant cultural tropes.[48] She was roundly excoriated by almost everyone, particularly by "real" African-Americans who felt she had duplicitously appropriated their identity. The defenestration of Dolezal is a particularly good example of the tendency toward authenticity in identity politics because, ironically, the NAACP was founded as a multi-racial organization—as it still formally is—where "blackness" was never a requirement for membership. There were potentially separable issues with Dolezal and her credibility, such as her claim to be a victim of racial harassment—the assessment of which presents quite the conceptual quagmire for identitarians—but it is not immediately obvious why her African-Americanness or lack thereof would be relevant to her leadership position in the NAACP. Though it is easy to see why her alleged dishonesty may have been.

Dolezal's performance of African-American identity had a long run but was finally shut down by her public exposure and then the judgment levied by the amorphous mass of self-appointed authenticity guardians (a search for "Dolezal" on Twitter will immediately yield them—they are legion).[48] The performance collapsed not because of the content of her activism or her going off script, but because of what seemed to be her personal qualities and, above all, when she was exposed as lacking proximity to what was considered by authoritative interpreters to be the baseline reality of the African-American experience.[49] This latter is unspecified but it seems to be comprised of a mixture of biology (which is odd given that race has been discredited as a useful biological concept)[50] and, probably more significantly, that she did not really fully *live* the African-American experience since she grew up white, had white parents and existentially had to know that *for her* the identity was voluntary rather than ascriptive—hence the duplicity, one deeper than merely lying to others. There is nothing inherently inconsistent in finding moral fault with Dolezal (though one wonders how doctrinal acceptance in many quarters of a Butlerian "performativity" standard regarding gender and sexual orientation is consistent with the judgmentalism). But I am not interested in moral appraisals of Dolezal's personal character. For present purposes what her case illustrates is how, when it is pushed, identity politics reveals its underlying authoritative-proximity structure, where legitimacy is preconditioned by *perceived proximity* to a constructed ideal of group authenticity that itself is perceived as not having been constructed (and is in fact taken to be more "real" for that reason.) Though it may appear at first to have an ostensible affinity, this strong tendency toward hierarchy does not square well with the left's defining egalitarian impulse.

In this manner, by their different routes, identitarianism in full flower ends up in the same political place as biblicalism (or any text-based religion): it erects an authoritarian hierarchy based on Truth proximity, identitarianism doing so as a function of authenticity and biblicalism as a function of its system of authorized interpretation. Doctrinal ambiguities and inconsistencies are actually enabling for such power structures as they provide growth opportunities for the ensuing priestly castes. The relevant proximities of power must be continually re-measured.

Owing to its founding premises, no identitarian ethno-state is likely to be much of a socialist paradise. Since it bases authority on segmentation, it will always need external and internal Others to assign propinquity; nearness to an imagined ideal of authenticity is its touchstone of moral value. In this respect there is a strangely Platonist structure to these identitarian conceptions, as they are reminiscent of the work of neo-Platonic philosopher Plotinus, where perfection is conceived as *proximity* to "the One." As with Plotinus, for identitarians reality admits of degrees as function of that proximity.[51] If *per impossibile* the ethno-paradise were created, it would continually have to manufacture enemies, a cozy identitarian version of Orwell's "Two Minutes Hate," where the Emmanuel Goldstein bogeymen would still need to be made objects of ritual obloquy for in-group cohesion.[52] Far from the securing of "social justice," if they are to be consistent, egalitarians would have to find this kind of "othering" to be dystopian. Identitarianism appears to be egalitarian when it surfaces on behalf of a historically oppressed group, but when it is examined closely its motivations are different. Its metaphysical comforts derive from group connectedness and a sense of *place*. The egalitarianism is in principle detachable from its worldview. More, it is *likely* eventually to be ditched as impediment to the internal hierarchies identitarians erect when they are left to their own devices. Absent a more holistic worldview anchorage, the identity focus tends to degenerate into an authenticity cult where moral membership is a function of proximity to an imagined ideal type and/or community. When this occurs it severs any recognizable connection with the egalitarian left.

Futurism

Aesthetic and technological futurisms present yet another available worldview category. These perspectives spring from innovations within a particular realm, in this case art or technology, where visionaries are no longer satisfied for their innovations to remain internal to their originating spheres of activity. The visionaries become convinced that their innovations are *more* than just artistic or technological advances and then take steps to consolidate them into a more comprehensive conception of the good. It is important, then, to distinguish largely intramural developments in particular areas from those that are expansive enough to count as offering broad-spectrum worldviews.

Pre-World War I Cubism, for example, associated with the early work of painters Pablo Picasso, Georges Braque and Diego Rivera, explored aspects of visual perception but did not possess a broad agenda to make over human life; it was pretty safely confined within the canvas. While in time it had a somewhat broader legacy, even then cubism's spillover influence remained mostly confined to the world of art, mainly sculpture and architecture.[53] Even a more diverse movement such as Bauhaus, with farther-reaching incursions into architecture and design, although associated with a more holistic modernizing sensibility, still operated largely within traditional aesthetic boundaries. As articulated by founder Walter Gropius, Bauhaus made its focus to "create a new guild of craftsmen" who would "unite every discipline, architecture and sculpture and painting."[54] (Despite these self-imposed limitations, the Nazis still designated Bauhaus as "degenerate art.")[55] Yet there have arisen occasional art movements with ambitions to become something much more and, despite their provenance in the art world, position themselves as blueprints for how to refabricate human life generally. A paradigmatic prototype might be Renaissance humanism, which was clearly a multi-disciplinary ideological project of almost boundless cultural ambition that the Church rightly came to see as deeply threatening. Contemporary examples include artistic movements of different political trajectories such as Italian futurism and Russian constructivism (e.g., Marinetti, Tatlin), the ideologically energetic Situationist International (associated with the Paris uprising of May 1968 and other counter-hegemonic trends) and, on a smaller scale, the playful American "anti-art" Fluxus community of the 1960s-1970s.[56]

Italian futurism, for example, was about far more than visual arts. The movement sought to redirect just about every area of Italian life as a madcap "insolent challenge to the stars" featuring a "love of danger" and "the beauty of speed," in the service of their notorious aim "to glorify war—the only cure for the world—militarism, patriotism, the destructive gesture of the anarchists, the beautiful ideas which kill, and contempt for woman."[57] As critic Robert Hughes describes it, futurism trumpeted itself "as a broad 'revolution' in living aiming to change life itself, embracing everything from architecture to athletics, politics and sex."[58] Consistent with this audacity, many in the futurist orbit embraced the surreal spectacles of reality show star *avant la lettre* Gabriele d'Annunzio, which included his 1919 freelance military invasion across the Adriatic to set up the "Free State of Fiume" in what is now Rijeka, Croatia. And they routinely ventured beyond traditional artistic endeavors in such initiatives as *Futurist Manifesto* author Filippo Tomasso Marinetti's quixotic campaign to have pasta banned (in Italy!) and his bizarre *Futurist Cookbook* with strange recipes involving *eau de cologne*, ball bearings and other unlikely ingredients, along with strange instructions for how to eat them, like while accompanied by a drumroll.[59] With a great deal of proto-fascist panache, the futurists refused limiting themselves to art.

In this they are unlikely conceptual bedfellows with the Situationists and Fluxus who, despite their obvious ideological differences, exhibited analogous gestures of refusal to stay within the conventional boundaries of art. For their part, the Situationists wanted to redefine art completely by taking it to "a higher stage," where "everyone will become an artist, i.e., inseparably a producer-consumer of total culture creation, which will help the rapid dissolution of the linear criteria of novelty."[60] And Fluxus's founder George Maciunas states, by de-commodifying artistic production and making it less consumer- and more communally-oriented, "Fluxus should become a way of life not a profession."[61] Though their substantive outlooks could not be more different, Italian futurism, Situationism and Fluxus all shared a similarly eudaimonistic breadth of ambition;

they aimed to enact, in Fluxian terms, *ways of life*: worldviews.

In addition to aesthetic futurism, broadly defined as art world-originating movements with relatively comprehensive extra-aesthetic aims—in this sense Renaissance humanism is a futurism vis-à-vis the quattrocento—there are more purely technological visions that are inherently futurist due to their strong reliance on extrapolations from existing technologies. Visions of this type include certain sci-tech futurisms such as computer scientist Ray Kurzweil's "singularity" and other speculative human-machine syntheses, including those involving drug-induced hallucinogenic mind-chemical mixtures. There is great popular interest in this area as evidenced by the popularity of films such as *Blade Runner* (based on Philip K. Dick's original vision)[62] and other cinematic "cyberpunk" depictions of cyborgs and human-machine amalgams or "biopunk," such as the BBC TV series *Orphan Black* about the moral and political dilemmas of genetic engineering and human cloning. Kurzweil's singularity anticipates a near future in which artificial intelligence far surpasses that of humans and we become fused in various ways with machines.

In a sense human beings have always been fused with technology in that they are inseparable from their surrounding ways of life, for example, techniques of agriculture giving rise to civilization itself or the famous instance of the stirrup generating feudalism (because it generated martial realities that necessitated the maintenance of mounted knights that in turn required the feudal socio-political infrastructure).[63] But singularitarians have something more comprehensive in mind, sometimes going as far as to suggest that a kind of immortality is ours for the taking via, in essence, downloading ourselves onto a new substrate in a glorious "bionic convergence."[64] They also remind us that there is no warrant other than an unjustifiable "carbon chauvinism" for eschewing alternative, perhaps electronic and silicone-based platforms for life. Furthermore, as scientist and inventor James

Lovelock, progenitor of Gaia theory (about whom more later), points out, "The only difference between non-living and living systems is in the scale of their intricacy, a distinction which fades all the time as the complexity and capacity of automated systems continues to evolve."[65] Lovelock adds, "there is no clear-cut rational explanation that distinguishes self-organizing inorganic systems from life."[66] In principle the way seems clear for any number of bio-tech fusions involving humans or other animals and inorganic material.

Whatever the ultimate plausibility of such a scenario, and one cannot help but be questioning of extreme versions of this vision due to their oddness (for example, downloading "ourselves" into our electronics), for present purposes what is significant is that the singularity, even if inchoate, seems to have the makings of a bona fide comprehensive conception of the good. Kurzweil explains:

> Although neither utopian or dystopian, this epoch will transform the concepts we rely on to give meaning to our lives, from our business models to the cycle of human life, including death itself. Understanding the Singularity will alter our perspective on the significance of our past and the ramifications for our future. To truly understand it inherently changes one's view of life in general and one's particular life.[67]

What precisely is to be the content of all of this new stuff is unspecified—which is why singularity is mostly inchoate as a worldview—but it does seem to have many of the requisite high and low elements. On the high end there is the cosmic, almost ascension-like aspiration toward bionic convergence that science writer John Horgan considers an "escapist" fantasy: "**Let's face it.** The singularity is a religious rather than a scientific vision."[68] In all fairness to singularity, though, Horgan should be reminded that the great world religions—certainly Christianity—are *literally* "escapist" too. And that fact has not stopped them from gaining billions of adherents. Escapism is in fact a necessary ingredient for their ideological success; when hearing the charge "escapist" one wonders what exactly is its opposite and whether that "anti-escapism" too might be overrated. In terms of everyday life, one might point to the increasing number of areas where we seem more and more intricately fused with our tech, from the metastasizing indispensability of smart phones and wearable tech to the high-tech home invasions currently being perpetrated by devices such as Amazon's Alexa and Google Home.

Yet focusing specifically on devices may miss the larger point. Consider the fact that as of 2017 online dating sites represent the second most popular origin for heterosexual couples and are number one by far for homosexuals (they also have arguably boosted interracial dating).[69] Thus even the most intimate aspects of life can be seen as functions of complex algorithms that are no longer really under the direction of any flesh and blood human beings. And this is not to mention the monumental levels of commercial and government surveillance underway from our favorite websites and apps for directing attention and behavior across the board. These quotidian human-tech fusions at the low end may actually be a conceptual strength of some sort of singularity-styled worldview. Yet one wonders what kind of morality is conceivable, as the ecstasies of the imagined bionic convergence seem not to leave much room for interpersonal care and concern. A counter-argument, I suppose, could be that if we were all to be biologically converged it would represent the ultimate in human interconnection—though it seems unknowable whether this would be pleasurable or maybe some kind of new hell. The desirability of its endgame would need to be developed, I suspect, for singularity to take off any kind of popular and durable worldview.

In sum, the techno-aesthetic futurist category tends to be populated with very small and highly intellectualized coteries of creative types who place their emphasis on grand visions rather than everyday life. Despite their ambitions—and some highly game if not quixotic attempts—futurisms therefore perpetually risk disconnection from ordinary people and they often devolve into relatively short-lived *avant-garde* fads, as in the case of some of the art movements. Their creative power can be intoxicating though (sometimes literally as in reportedly life-altering ayahuasca/DMT and psilocybin trips) and one can expect globalism and online interconnectivity to augment these movements' reaches.[70]

Worldview fusions

Enticing and suggestive as they are, these visions are mostly embryonic as worldviews. Perhaps their ultimate value might lie in their potential role in the provision of important fusible ideological elements. And indeed there are plenty of cross-category fusions as well as individual streams internal to the categories that I have not named. As worldviews are complicated and any attempt to categorize them is merely a heuristic and hence artificial, one should remember that, just like religions, worldviews of any sort are *always* fusions of many elements—culture- and history-bound creatures that we are, there is no other possibility. A shared metaphor from biology and geography is helpful here: *anastomosis*. In any root-like or brachiated system such as the roots of a tree, intestines, lungs, arteries and capillaries in the circulatory system, a river system of tributaries or streams, there are sometimes instances where the roots or branches cross over one another and knot up; they have their own sources but they can become intertwined later on such that they form their own loci. This could even be intentional as in medicine where intestines could be sewn up together to re-establish functioning after colon cancer or naturally occurring as when tree roots or streams crisscross and converge with one another. When anastomoses form like this, although their parentage is elsewhere, they can take on their own identities as distinct entities in their own right.

As the archetypal narratives by which human beings have made sense of ourselves, worldviews in the ensemble are like a vast forest with an even vaster root system backward in time, where roots growing from one part of the forest can always anastomose with those growing from another. Surprising anastomotic fusions are always possible in this way. Those looking for optimism might find in all this collective connectibility a powerful set of latent creative forces.

For instance, one off-the-beaten-path worldview anastomosis is found in the currents of thought comprising Afrofuturism, which, as the label indicates, fuses Afrocentrism with futurism, and is a relatively recent formation with a particular emphasis on science fiction. Among its inspirations are the works of science fiction novelist Octavia Butler and the works of musicians such as Sun Ra, George Clinton/Parliament and, on the contemporary scene, Janelle Monáe, that envision reconstructed utopias against the backdrop of African-American cultural loss and diaspora. Even more recently, there is "Afrocentrism 2.0" and the "Black Speculative Arts Movement (BSAM)," "an embryonic movement examining the overlap between race, art, science and design."[71] Afrocentrism 2.0 goes beyond identity politics as usual and is far more imaginative and forthrightly eudaimonistic. In the words of communications scholar and curator Reynaldo Anderson:

[T]he early twenty-first century technogenesis of Black identity reflecting counter histories, hacking and or appropriating the influence of network software, database logic, cultural analytics, deep remixability, neurosciences, enhancement and augmentation, gender fluidity, posthuman possibility, the speculative sphere, with transdisciplinary applications...

has grown into an important Diasporic techno-cultural Pan African movement.[72]

As represented by a broad range of creative output, Afrocentrism 2.0 and BSAM scrupulously avoid eurocentrism yet move beyond mere complaint and critique by seeking to integrate "African diasporic or African metaphysics with science or technology and seeks to interpret, engage, design, or alter reality for the re-imagination of the past, the contested present, and as a catalyst for the future."[73] An example of the project may be found in the 2016 New York Public Library exhibit *Unveiling Visions: The Alchemy of the Black Imagination,* curated by Anderson and artist John Jennings, which art historian Tiffany Barber describes as representing "an emergent strand of black cultural production that combines science fiction elements to imagine alternative visions—sometimes reparative, sometimes not—of the black experience in the past, present, and future."[74] As I have been arguing this willingness to "imagine alternative visions" should be regarded as the golden chalice of contemporary cultural production.

Another interesting fusion along these lines might be found among the ayahuasca churches emerging in the United States, accompanied by lengthy legal battles, that synthesize chemical technologies like DMT (the active ingredient in ayahuasca, a plant-based brew) and nature-centered enframing ritual patterns imported from certain shamanistic indigenous religions of the Upper Amazon.[75] The fusion here is between modern technology in the form of chemistry (uncovering knowledge about the relevant properties of these powerful drugs) and appreciating the archaic indigenous, usually shamanistic, contexts in which these plant-based hallucinogenic experiences have been practiced for millennia. A further empirical point in favor of the hallucinogenic route are the fascinating studies by medical researchers at Johns Hopkins University of psilocybin mushrooms and their notable effectiveness at diminishing the fear of death among the terminally ill and which seemed to help study participants find greater optimism and peace in their lives.[76] One of the main hallmarks of a bona fide worldview is its ability, if not fully to answer, to at least reconcile people with their mortality. This element *alone* makes these chemicals essential to examine for anyone seeking to re-establish narratives of meaning in the contemporary world. Given their potency it is unsurprising that these drugs are perhaps best approached via the ritual contexts of indigenous religions and those traditions' accumulated knowledge about handling and administering them. It would be breathtakingly arrogant for us to assume that we have nothing to learn from ancient practices such as these that have been central to the human experience for thousands — and maybe hundreds of thousands — of years on every part of the globe.

I mention these examples to emphasize that my four categories should not be conceived as heuristic rather than sharply distinguishable in any ultimate sense. As I indicated, they should be considered as a small copse made up of four big trees, nourished through extensive and entangled (and intercommunicative) root systems, with saplings of various sizes perpetually growing up around them.[77] Like the trees, though, the four approaches may usefully be *conceived* as separable too. My contention is that a convincingly durable left political program would need to root itself in some version of one of these worldviews or some anastomosed fusion across them like Afrofuturism or ayahuascan religion.

There is an unavoidable caveat though: where exactly views like these might finally take one, no one can know; there is an inherent element of contingency and *moral danger* here where *a priori* commitments cannot be guaranteed. One should expect to be ideologically *altered*.

People often speak of the need for "transformation" but to

take that imperative seriously involves appreciating two things:

1) as argued previously, the liquid modernity authorizing neoliberalism is already *deep into the process of transforming us* and our sense of selfhood by smashing the supposedly solid identities of the past and setting us—for better or worse—existentially adrift, and

2) *the only way out is through,* which implies accepting a prospective contingency where there is no way to ensure full compatibility with any existing political ideology. At its best we are talking about a kind of controlled demolition of the erstwhile self, which necessarily cannot fully visualize its successor, any more than a new parent can actually *see* who their newborn baby will eventually become. One might be able to start detecting the birth pangs, though.

At the political left's high-end structural level we have the purely negative "critique of neoliberalism" and at the low-end personal level we have the purely antagonistic call-out culture of identity politics. These have largely achieved their destructive pedagogical purpose and are now approaching their limits and starting to transform into something else, as exemplified in the petri dish of campus identity politics where the cutting edge is now to denounce basic civil liberties as mere shields for fascists, racists and rapists.

From a bird's-eye dialectical view, one can applaud this pressing forward because it lends urgency to the assessment of this "something else" while it simultaneously exposes the inadequacy of bloodless liberal proceduralism—as if calm compliance with legal norms of due process and equal protection solves everything. The salutary aspect of these developments in their ensemble is that they have unexpectedly pushed things to the point where philosophical discussions of larger institutional aims and purposes are becoming less avoidable.[78]

As odd as this may seem, and against our jaded relativistic impulse to rebel against even a descriptive moralism, what is odder still is to think that the contemporary left's thin morality is sustainable *as is*: a purely negative force with no substantive positive vision. Even "socialism" will get one only so far, even in its most attractively comprehensive form: a social movement unionism that gestures beyond deal-cutting for its own membership toward broader solidaristic ideals. Despite its urgency, such an approach remains a mere ad hoc aggregation of policy prescriptions that lacks an underlying account of its own moral commitments. It tends in an overly facile manner to take its humanist egalitarianism to be self-evident and so sees itself as excused from associating with a richer and more comprehensive eudaimonistic vision of the good. One should join up with the union because it is in one's enlightened self-interest to do so for solid material reasons. Yet the (weakly) promised solidaristic life is never given *content*—that discussion is always deferred, the urgency of the struggle at hand always taking priority. In fact, the old Marxist picture possesses an ironically laissez-faire notion of solidarity: a mirror image of the free market, their shared material interests against the bosses will, as if by an invisible hand, organically bring workers together via their unions and forge them into class-conscious revolutionary agents. They will then see that they need to seize the means of production, etc. and in doing so will create, well, some new *something*.

But perhaps something beyond a constellation of shared material interests is needed in order to forge a better world, namely, a more vivid *conception* of that world. The "collection of interests" starting point for solidarity is lacking something, as evidenced by the fact that, in American labor history, after a union wins it typically quickly deflates itself and reverts to labor peace and business-as-usual. This is only natural: people have to get on with their lives. But neither is it wholly surprising when *α'. Popular Front + Cultural end*

α. Cultural Front

the underlying rationale for unions' existence is enlightened self-interest. In traditional Marxism "rational self-interest" is not to be understood in moralistic terms, though, but it is all part of the plan, so to speak; the real oppositional work is to foster an *educational expansion* of that self-interest that will not be able to sit still once it has discovered itself as class-conscious. But in most MEDCs the dream of "social movement unionism" breaking out into revolution-winning solidarities has never really happened on a large scale.[79] This is not to say that movement along this spectrum is impossible—and there are inspiring local struggles against eliminationist austerity capable of creating larger ripples such as the successful 2018 West Virginia teachers' wildcat strike that inspired a wave of similar teacher strikes in other conservative states.[80] Commenting on West Virginia, legal scholar and social critic Jedediah Purdy memorably expresses this potential: "The teachers' movement is a reclamation and redirection of a militant working-class identity. They look back to the miners, look around at the uncompensated wreckage of the land, and look forward to the world we are all entering, where the labor is in social reproduction: teaching, caregiving, the upholding of the human world."[81]

The challenge, then, is to create real staying power by providing more content to Purdy's affecting phrase "upholding of the human world." An optimistic way to put it would be that social unionism consists of dots of material interest that are waiting to be connected by a larger normative framework that is not itself "issue-oriented."

But what might this connector be? Socialism and communism have been history's main candidates for this role. The latter had more comprehensiveness and vision but it is probably fatally discredited in the MEDCs. There are however promising signs for socialism among the rising generation.[82] But it is unlikely that keener awareness of material self-interest *alone* will create sufficient ideological buy-in. Reversing the traditional formulation, perhaps a more comprehensive vision *at the outset* might make more sense, especially in a liquid era where traditional workplaces are dissolving and would-be workers are left more and more on the outside looking in—and are, in fact, more and more desperate to reconstruct coherent narratives about themselves. Identity proliferation is in this sense both the challenge and the opportunity: under conditions of chronic precarity and fragmentation, it is an against-the-odds undertaking to turn everyone's attention toward some shared framework of meaning. Unionism is just not built for this job; it is good in a fight but tends to lose coherence and purpose when the dust settles. Like other MEDC left formations of the last century, it is better at *militancy*—the noble fight *against* exploitation, cruelty, greed, poor working conditions—than it is at turning the corner toward providing a sustaining vision for the long haul.

This omission is not unique to social movement unionism. We have arguably reached a point of *peak critique* of neoliberalism where the growing piles of essays and books matter very little (my own included) if along the way a more holistic and *nourishing* set of orienting values does not begin to emerge. While philosophy is limited in its ability to provide what is needed (as is any single approach) it can help gather and synthesize and gesture toward possibilities. To do so, it has to be willing to look bad, though, and cut somewhat against the grain of modern academic specialization. The specialized "staying in one's lane" mentality that has rendered university-based humanities types timid and self-referential, pretending to advance their fields on the remunerative model of the natural sciences; we have exchanged potential broader cultural influence for intra-academic position and prestige—careerism in a word. There is an argument to be made for a salutary narrowness, i.e., for the most painstaking and arcane scholarship, but the arts and humanities would be better served by reconstructing their ability to contribute toward an integrative endeavor that is not only "interdisciplinary" in the marketing buzzword sense of on-campus inter-departmentalism but, more dangerously, also connective with *extra*-academic streams of thought and practice. On its own, scholarly lucubration represents far too limited a range of experience.

Q' Is this tone somehow an echo of what Weber is describing in Science as a Vocation?

Chapter 5

A world made of Earth

The intrusion of this type of transcendence, which I am calling Gaia, makes a major unknown, which is here to stay, exist at the heart of our lives. This is perhaps what is most difficult to conceptualize: no future can be foreseen in which she will give back to us the liberty of ignoring her. It is not a matter of a "bad moment that will pass," followed by any kind of happy ending—in the shoddy sense of a "problem solved." We are no longer authorized to forget her.
Isabelle Stengers (2015)[1]

The intrusion of Gaia

Back to the future even further, the last grand option is a contemporary revival of archaic traditions associated with nature-worshiping worldviews such as forms of paganism, Druidism, Wiccan, various indigenous religions and the like. In the West these are mostly pre-Christian forms constituting subaltern traditions that have survived often under severe persecution. Though some trace lines of descent from ancient times, many were revived in the modern era, within the last few centuries, and have had interesting overlaps with secular political formations (e.g., Wiccan and feminism, Druidism and environmentalism). A unifying aspect of this catch-all category is a strong tendency toward robust identification with the natural world and a concomitant implicit critique, in the contemporary context of environmental crisis, of hegemonic Christian and technological/capitalist conceptions of human beings' rightful dominion over the Earth. There is also a vague "New Age" pop cultural aspect associated with these formations that has been present for a couple of generations now, with crystals, Tarot, alternative medicines and other assorted "spiritual" accoutrements present in many Western homes.

A tremendous advantage of this general tendency is that although wildly heterogeneous, the core suspicion of technological and industrial development and the generally-shared ecological sensibilities dovetail well with what is surely humanity's greatest problem: the possibility of pollution-driven catastrophic climate change and related issues such as the mass extinction of species and the rapid depletion of essential resources (e.g., oil, water, soil, metals). There is also a tendency to de-center and de-hierarchize our conception of the cosmos by recognizing nonhuman animals and plants—and even other entities—as loci of moral worth, wholly antagonistic to biblical conceptions of a divine right vis-à-vis nature and the stubborn moral legacy of the Aristotelian *scala naturae* that places human beings at the pinnacle of creation as the *telos* toward which other organisms allegedly strive.

I suggest that this general approach provides the most promising moral backdrop for left egalitarianism and is to be found in something like a resolute continuation of the Nietzschean extrapolation described earlier. It can also be thought of as admitting of several synthetic visionary combinations of the preceding worldview candidates. Actually, it is perhaps *best* thought of in this fashion. Like ancient paganism it looks toward a robust identification with the natural world first, yet like the various futurisms and singularities it is sanguinary vis-à-vis the idea that human possibility can play a redemptive role. In this respect it is the macrocosm to the microcosm conceived by the practices of organic farming and permaculture (described below). As a continuation of the momentum of the moral ontology of Jerusalem-Athens, it is also consistent with key elements of the biblical framework that has shaped Western culture—even as it seeks to reawaken older and much longer-lived moral impulses from pre-Christian times. It even takes something from the psychological vectors of identity politics. The idea here is to allow for an extra-anthropomorphic extension of, not only concern with the natural world but a radical intensification of our ability to *identify* with it, a phase shift of moral consciousness such that we see ourselves viscerally as component parts of a system (or systems) of life on Earth.

Ideally, this identification would simultaneously go high, in the previously-discussed sense, that is, appeal honestly to our rational faculties and be dynamic enough to mesh adequately with the natural sciences as they develop; it must convince on the transparent and open-source level of scientific inquiry rather than necessitating the esoteric and interpretive hierarchies that are borne out of doctrinal ambiguity and implausibility. Yet at the same time it must also go low (so to speak) and be capable of eliciting the imaginative and emotional responses that can pattern everyday life and help knit people together by something other than their cognitive faculties alone. It has to hit people where they *live and love* not just where they think. As such it must also be capable of providing powerful transformative experiences like those of religious conversions or, analogously, various hallucinogenic visions that have proven to be capable of radically altering individuals' outlooks and worldviews.

All this is a tall order. While it admits of infinitely many different approaches and perspectives, what I think could fill it are various permutations of what has been developed as "the Gaia hypothesis" by atmospheric scientist and inventor James Lovelock and microbiologist Lynn Margulis.[2] Stated in its starkest terms, the basic idea is that *the Earth itself is a living organism* or, at least, possesses at its macro-level a sufficient number of the characteristics of living organisms that it can be plausibly regarded as such. There are stronger and weaker ways to state the Gaia idea. But the core notion is that the Earth, like a juggler balancing many pie plates, can be seen to maintain a multi-faceted and self-regulating homeostasis; the "entire surface of the Earth including life is a self-regulating entity."[3] These processes occur on several different levels and across several different biosphere circuits that maintain improbable equilibria: oxygen and CO_2 levels, oceanic salinity and pH, and global surface temperature. Scientists are also studying weirder homeostatic phenomena such as the occurrence of mineral diversity and mineral evolution (both greatly augmented by living organisms)[4] and lithospheric interactions with organic processes such as photosynthesis (oxygenating the atmosphere sustainably).[5] Lovelock even speculates at one point that "migratory birds and fish serve the larger Gaian purpose of phosphorous recycling."[6] (In later writings he becomes far more cautious about using terms such as "purpose" due to their teleological implications that Gaia has goals and intentions like we feel ourselves to have as persons.) In these and other areas, biotic and abiotic entities are engaged in mutually reinforcing and symbiotic feedback loops where mutualistic and emergent Gaian networks function to "maintain a habitable environment for whatever is its biosphere."[7] Gaia might be described as a system of systems, "a complex entity involving the Earth's biosphere, atmosphere, oceans, and soil; the totality constituting a feedback or cybernetic system which seeks an optimal physical and chemical environment for life on this planet."[8]

Atmospheric oxygen is a good example. If it were much higher than 21% then the world would be so combustible that forest fires and the like would be catastrophic; yet if it were too low then of course oxygen-breathing animals would not be able to survive. All the while, however, plants and animals (with crucial assists from atmospheric layers) maintain via their exhalations the appropriate levels of oxygen and CO_2 for one another. Astrobiologist David Grinspoon elaborates:

All this oxygen we take for granted is the byproduct of life intervening in our planet's geochemical cycles: harvesting solar energy to split water molecules, keeping the hydrogen atoms and reacting them with CO_2 to make organic food and body parts, but spitting the oxygen back out. In Earth's upper atmosphere some of this oxygen, under the influence of ultraviolet light, is transformed into ozone, O_3, which shields Earth's surface from deadly ultraviolet, making the land surface habitable. When it appeared, this shield allowed life to leave the ocean and the continents to become green with forests. That's right: It was life that rendered the once deadly continents habitable for life.[9]

Seeming as if by magic, the same homeostasis can be observed with climate temperature. Though heat from the sun has varied enormously over billions of years, the Earth's systems have perpetually adjusted to keep things within a narrow life-conducive range.[10] Absent these homeostatic mechanisms, it would have been probable that the Earth would have just "died" and become permanently uninhabitable like the other planets in our solar system. There is as one might expect much debate among scientists about every aspect of these mechanisms, including over whether or not certain ways of speaking about Gaia are ultimately appropriate or helpful at all.[11] This includes those who dislike the term itself because of its Seventies "New Age" connotations and the tendency of some Gaia enthusiasts excessively to poetize and personify it as "Mother Nature" and, ironically, to anthropomorphize it as some kind of inherently benevolent guardian. In response to such unscientific excesses some researchers prefer the more buttoned-up label "earth system science." As Margulis insists, "I cannot stress strongly enough that Gaia is not a single organism. My Gaia is no vague, quaint notion of a mother Earth who nurtures us. The Gaia hypothesis is science. The surface of the planet, Gaia theory posits, behaves as a physiological system in certain limited ways."[12] With undue spiritualism stripped away, and by whatever name these organic-inorganic symbiotic systems are called, it is clear by now that there is a high level of plausibility to at least *some* version of the Gaia perspective such that one does not need to take it as *merely* mythical or poetic. No leap of faith is needed; "The Gaia hypothesis is science."[13] There seems little doubt that Lovelock and Margulis's original Gaian vision of the Earth as a kind of superorganism—"the largest living creature on Earth"[14]—consisting of a complex simultaneity of complementary self-regulating systems, is intellectually powerful and rationally convincing as a fecund *perspective* on our planet, ourselves and our interrelation.

Beyond the science—and here is where one must tread very carefully—it also bears all the hallmarks of a psychologically compelling and historically available worldview that extrapolates in the Nietzschean manner from our current moral and legal egalitarian universalist ideals toward recognition and identification with this larger and more encompassing Gaian system of systems. Rightly tying things back to Kant, philosopher Bruno Latour explains that "We are faced with the 'generalized revolts of the means': no entity—whale, river, climate, earthworm, tree, calf, cow, pig, brood—agrees any longer to be treated 'simply as a means' but insists on being treated 'always also as an end.'"[15] In other words, contemporary people are starting to discover—sometimes in delight, sometimes in horror—that it is not good enough to extend care and concern and even equal treatment only to other human beings (as difficult as *that* universalist move continually proves, for example, the persistence of racism) but also to entities that have traditionally been considered to be mere things and as such outside legitimate moral concern. Not only are there widespread sentiments for such as animal rights and concern for future generations but many of us also are able to experience suffering and a real sense of loss at a coal company blowing the top off a mountain or the logging of an old growth forest or a destruction of remnant wilderness—not to mention the extinction of a species.

This is the moral-psychological heart of what I have been calling the Nietzschean extrapolation: the idea that not only *people like me* (tribe, soul, humanity) count and are worthy of moral consideration, but also other nonhuman sentient beings and, ultimately, even certain inorganic entities, including those at a large scale, what Timothy Morton calls "hyperobjects."[16] A forest or an ocean or a mountain chain would count as hyperobjects. A spectacular recent example of the extension of consideration to hyperobjects is New Zealand's decision to grant 120,000-year-old Mount Taranaki, a dormant volcano on the west coast of North Island, "the same legal rights as a person"—in fact the third "geographic feature in the country to be granted a 'legal personality.'"[17] Along these lines, Latour's point is that it is no longer acceptable to treat these heretofore *mere things* with the customary moral obliviousness; because we have started to feel their claims, contra centuries of Cartesian mechanism, they can no longer be consigned to mere thing-hood. It is now possible to do *wrong* by them in a way that it has not been for a very long time—though this has never stopped being possible for *every* indigenous culture that remains around and within us. As philosopher Mary Midgley appropriately observes, the "current Gaian thinking that I believe can help here is a new scientific development of an old concept. The imaginative vision behind it, the idea of our planet as in some sense a single organism, is very old. Plato called the Earth 'a single great living creature' and this is language that people in many cultures would find natural."[18] As a case in point, the effort on Mount Taranaki's behalf was pushed by the local Maori tribes, who view the mountain "as an ancestor and whanau, or family member."[19] It is therefore perhaps more accurate to say that many of us are *re-discovering* something very ancient with the Gaia idea rather than coming up with a novel conceptual innovation.

Still, the rediscovery is occurring in our present context and so we will see it according to our own assumptions and pressing concerns, especially the ongoing environmental catastrophe—to be sure a catastrophe for *us* and not necessarily for resilient nonhuman Gaia as a whole. So the question is not our doing any favors for Gaia, which is not really possible. As Margulis strongly emphasizes, "We cannot put an end to nature; we can only pose a threat to ourselves. The notion that we can destroy all life, including bacteria thriving in water tanks of nuclear plants or boiling hot vents, is ludicrous."[20] So we need to take Gaia into account not out of altruism but for *ourselves*, recognizing that one of the deepest lessons in doing so is that the conceptual membrane that once, in retrospect unjustifiably, separated "human" from "animal" is just as illusory ultimately when it comes to even deeper dualisms like "organism" and "environment" and even "biotic" and "abiotic."

With this radical permeability in mind, Latour introduces his strange notion of a "parliament of things" pursuant to an "object-oriented democracy" where we would make some attempt at allowing nonhuman "voices" to be heard, where a Gaian Earth is presumably to be accounted among the recognized "members." [21] I confess that I cannot make detailed sense of what Latour might mean here. But I am able to embrace what I take to be his motivating imperative, which is that we begin somehow to take into account the claims that these newly-admitted (to our own moral regard) Others seem to be making with such insistence, in the ensemble the "intrusion of Gaia," in philosopher Isabelle Stengers' memorable phrase.[22] The long view from this perspective, what Spinoza termed *sub specie aeternitatis* (Lat. "under the aspect of eternity") is beautifully captured by Grinspoon as he reflects on Gaia's meaning for our own self-understanding:

This gives us a different way to think about ourselves. The scientific revolution has revealed us, as individuals, to be incredibly tiny and ephemeral, and our entire existence, not just as individuals but also even as a species, to be brief and insubstantial against the larger temporal backdrop of cosmic evolution. If, however, we choose to identify with the biosphere, then we, Gaia, have been here for quite some time, for perhaps 3 billion years in a universe that seems to be about 13 billion years old. We've been alive for a quarter of all time. That's something.[23]

The sense of smallness may seem at first deflating; it might be taken to diminish the significance of our best-laid plans and largest-scale undertakings. Yet it provides the key to how the Gaia perspective can fulfill one of a worldview's most important jobs: to provide comfort regarding mortality. In the longest span, of course, even the planet comes to an end but I suspect that Gaia, as a dynamic system of systems, is capable of opening out beyond the physical planet as well. Grinspoon may place too much emphasis on linear duration in the above quote, but the enlarging Spinozan sensibility toward which he gestures is the main point.

The key is our ability to make that initial identification, one that takes us beyond, certainly, the cramped individualism of liberal modernity and also the universalist way station that asks us to see ourselves as part of humanity and so focuses us on what makes us as human beings *distinguishable and separate* from other living—and certainly nonliving—beings. It is morally more defensible to identity with a common humanity than just a part of it, say, than being a nationalist, racist, sexist, ableist or whatever exclusions are dictated by the moral lacunae of one's particular setting. But the extrapolative momentum—a morally expansive force that can be motored by a Gaian sensibility— is capable of pushing one farther into a more comprehensive identification with the systems of life. Like the sensation of consciousness itself, the sheer fact that this identification can be *felt* is good enough for it to be considered as having existence; as with consciousness there is no worry about being fooled because my very capability of being fooled is sufficient proof that I am conscious—if only as *that-which-can-be-fooled*. Despite the antipathy toward Descartes for disanimating the natural world, one should remember that he was deeply insightful about some of the most important matters, in this case the basic argument of his famous *cogito*: "I think therefore I am."[24] If this kind of identification can be felt, therefore it *is*.

This is the point where I think argumentation breaks off and another kind of discourse takes over. Poets, who typically operate at the edge of ineffability, are more likely better equipped to articulate what I am getting at here, which is that sense of being able to derive what philosopher Richard Rorty called a "metaphysical comfort" from attaching oneself to the larger living world.[25] Two powerful examples are, first, the classic poem by the Roman Stoic Lucretius, *On Nature*, the point of which is to forge this very intimate identification under the shadow of mortality and to allow us to conceive ourselves as inevitably returning to a wider cycle of life:

> And part of the soil is called to wash away
> In storms and streams shave close and gnaw the rocks.
> Besides, whatever the earth feeds and grows
> Is restored to earth. And since she surely is
> The womb of all things and their common grave,
> Earth must dwindle, you see and take on growth again.[26]

A second example is found in the epigraph to this chapter and in the searing poems of Robinson Jeffers generally, as when he suggests, quite on point, that:

> We must uncenter our minds from ourselves;
> We must unhumanize our views a little, and become confident
> As the rock and ocean that we were made from.[27]

Or on the occasion of an encounter with a vulture, Jeffers allows himself to *feel*:

I tell you solemnly
That I was sorry to have disappointed him. To be eaten by that
beak and become part of him, to share those wings and those
eyes — What a sublime end of one's body, what an enskyment;
what a life after death.[28]

These sentiments are either heavenly or hellish; though I suppose
from the perspective of Gaian life-identification they are properly
understood as *neither*. Lucretius and Jeffers, I think, capture
something of how a gestalt shift toward Gaia might actually be
experienced from the inside, its qualia—the way it might feel.
From the perspective of the (incoherent) personal immortality
promised in the biblical traditions, it seems strange and gloomy
at first, this talk of being "restored" and "dwindling"—and still
worse, "unhumanizing" and "enskyment." At least it seems that
way. Until it no longer does. Why should we *not* identify with
these things?

Feral activists

She would turn her back on the modern world with the telegraph
and the railroads and its elaborate political constructions piled
layer upon layer. Everything gone...everything. And she would
live in constant movement over the face of the earth, in gratitude to
the sun and the grass, often dirty and lousy and wet and cold like
those on the other side but she did not care.
Paulette Jiles (2016)[29]

However compelling, though, is it possible to derive ethical
and political commitments out of such abstract and cosmic
ruminations? The Nietzschean extrapolation promises a follow-
through specifically of *moral* commitments based on an expansion
of the substrate to which those commitments adhere (i.e. soul,
subject, life, etc.). What might it look like placed in a more concrete
setting? Is it possible to *live* such commitments in the social world

rather than their being purely epiphanous ideations, as powerful as they may be, merely fragmented and momentary aesthetic glimpses of a worldview that is still far removed from us?

For clues, I would look away from all the cleverness and instead at a stand of old growth trees located in Mount Hood National Forest, Oregon in 2000. As described by religion scholar Sarah Pike in her brilliant study *For the Wild: Ritual and Commitment in Radical Eco-Activism*, literally *in* those trees was an ongoing militant environmental action, a "tree sit," engaged in by young environmental activists (they are almost always very young) to protect from logging area forests and their nonhuman animal inhabitants:

> High above the forest floor, activists had constructed a platform made of rope and plywood where several of them swung from hammocks. Seventeen-year old Emma Murphy-Ellis held off law enforcement for almost eight hours by placing a noose around her neck and threatening to hang herself if they came too close. Murphy-Ellis, going by her forest name Usnea, explained her motivation in the following way: "I state without fear—but with the hope of rallying our collective courage— that I support radical actions. I support tools like industrial sabotage, monkey-wrenching machinery and strategic arson. The Earth's situation is dire. If other methods are not enough, we must not allow concerns about property rights to stop us from protecting the land, sea and air." [30]

Pike goes on to explain that "Murphy-Elis speaks for most radical activists who are ready to put their bodies on the line to defend trees or animals, other lives that they value as much as their own." [31] This is but one of innumerable actions where these youthful activists physically endanger themselves—often risking serious injury (some have been killed) and long jail sentences—in order to follow what they feel is a moral imperative toward nonhuman entities such as animals, forests and wild places in general, "for the wild" being a terse salutation and slogan that they often use.

Their sincerity and commitment is made abundantly evident in the annals of their actions. One must really sit back and imagine a *17-year-old* with a noose around her neck, knowing that activists just like her have been severely injured and killed in these intentionally-induced states of "utter vulnerability." The latter phrase is apt because, in putting themselves on the line like this, these activists are voluntarily choosing to recreate, using their own bodies, what theologian Andrew Linzey suggests is the "utter vulnerability" of the nonhuman entities they are endangering themselves in order to protect.[32] (Linzey argues that the inherently passive state of the victim in cases such as these renders our moral responsibility toward them *more* urgent than it would otherwise be.) Or the devotional mindset possessed by activist Julia Butterfly Hill ("Butterfly" is another forest name) who famously conducted a tree-sit in "an ancient 180-foot California redwood called Luna" on a platform in the late 1990s *for 738 days*.[33] She ultimately saved the tree. Or contemplating the many others who have been jailed or injured or killed. Again: it is not a matter of agreeing or disagreeing with their cause or methods. The point is how demonstrably seriously they take their cause and the reverent commitment with which they pursue it.

By contrast it is just impossible to imagine, despite their episodic volume, anything comparable in terms of life-or-death intensity among the participants in twitter shaming mobs or campus call-out culture; as they are grounded in very little besides the flickering of situational emotions, these kinds of activities seem light and unserious by comparison. Activists such as Julia Butterfly Hill and Emma Usnea Murphy-Ellis not only *voice* their views, they come to *embody* them. Echoing the Martin Luther King Jr maxim, their cause seems so much more *alive* via their willingness to sacrifice themselves. Martyrdom can of course be used and abused and it can be weaponized by terrorists and other undesirables. But absent mental illness, it must always be taken seriously for, like it or not, it is a gesture that has always possessed tremendous cultural power. All of us armchair rationalists should allow the image to sink in of that 17-year-old, in a remote forest, without glitz or glamor, without TV cameras or status updates, placing that noose around her own neck on a wobbly platform in order to stop the trees around her from being felled. It seems insane and inspiring all at once, and it in a way creates a distance between her and others of less intensity, as if she were an alien from some moral planet the rest of us have perhaps glimpsed but not yet inhabited. And that is just it: she *is* at that moment an alien from another moral world, and she is demanding that we either join her in it or gaze at her from our own. Or, just as likely, merely be puzzled by the spectacle of someone who unapologetically *has* a worldview at all.

To my mind, whatever the debate about tactics (predictably the ranks are divided), these activists represent a sort of cutting edge of moral consciousness and are a concrete example of what I have been calling the Nietzschean extrapolation and one iteration of what Bruno Latour has in mind with his parliament of things idea. This is a powerful instance in which the ancestral egalitarian ethos is being expanded yet again, from a humans-only universalism to one that is *viscerally* biocentric. Pike's assessments are along these same lines. She elaborates that:

[t]hese activists are a radical wing of a broader cultural shift in understanding humans' place in a multispecies world and a planet in peril. Their actions express trends in contemporary American spiritual expression and moral duties to the nonhuman at the turn of the millennium. . .that decenters the human and calls for rethinking our appropriate place in the world vis-à-vis other species.[34]

196

There is a further deep mimetic identification that these activists undergo with nonhuman life where for themselves and others they "highlight the arbitrariness of the species boundary and remake their kinship with other animals"[35] and in so doing enact and experience "the blurring of boundaries between human and nonhuman bodies, so that self and other are no longer directly separate beings."[36] As Pike describes them, these forest and animal rights activists undergo profound and wrenching personal transformations—their very bodies are transfigured in various ways—every bit as consuming and life-altering as born again evangelicism and the many other types of Road to Damascus conversion narratives that populate the annals of Christianity.

In his memoir *How I became an Eco-Warrior*, Jeff Free Luers relates his own tree-sit turning point in a tree called "Happy":

I sat cross-legged, my back against Happy, and I began to meditate. I forgot that there was a plywood platform below me. I forgot that I was a single entity. I felt the roots of Happy like they were my own. I breathed the air like it was part of me. I felt connected to everything around me. I reached out to Momma Earth and felt her take my hand. I could feel the flow of life around me. I felt so in tune with the ebb and flow of the natural cycles. I asked Her what it was like to have humanity forget so much, and attack her every day like a cancer. I told her I needed to know, I needed to feel it. . .She granted my request. My body began to pour sweat. I felt the most severe pain all over, spasms wracked my body. Tears ran down my face. . .I could feel every awful thing our "civilized" way of life inflicts on the natural world. The feeling lasted only a second, but it will stay with me the rest of my life. . .My life changed that day. I made a vow to give my life to the struggle for freedom and liberation, for all life, human, animal, and earth. We are all interconnected, we are all made of the same living matter, and we all call this planet home.[37]

I see the emergent moral sensibilities of these young environmental activists as providing tangible elements of what an evolving Gaian ethic might actually *feel* like. They also show what it is like for moral commitments to be based in a worldview that is actually heartfelt, even amidst our cynical and ideologically fragmented age.

This is not to say that these young activists are completely correct about everything, or that the certainty and absolutism that often marks their views and tactics is always defensible. As they themselves often realize in retrospect they are not. I would put it this way: they are not yet necessarily *mature* in their views; as Pike shows, in the "afterlife" of their activist years, most of these youths *grow in* to their views (I would not say "grow *out*"), that is, they find ways to elongate their commitments so that they are durable through the normal human life span.[38] They *life-size* their devotion one might say. Luers himself, who was sentenced to 22 years in prison (and served 9.5) for "ecotage" involving arson of three vehicles at a Eugene, Oregon SUV dealership, earned a college degree in landscape architecture and continues to work for forest restoration and by all accounts now lives a "normal" life with his own family etc. Yet he and others clearly continue to be devoted to their worldview. Luers says looking back that:

> I was a young kid, just turned 21. I went out, did a pretty small little action and got hammered with 22 years. But I have continued to be passionate about why I did what I did, and I think that resonates with people. . .the things that I'm struggling for are so utopian they seem almost ridiculous. Yet people want a fraction of that idealism in their lives.[39]

Neither does it seem to be at all the case in their activist "after-life" that Luers and most of his cohorts renounce their commitments and their fallen comrades. In fact it would be more accurate to say that they sacralize their dead through ritual commemoration and continue to work on behalf of "eco-prisoners" through such groups as the Earth Liberation Prisoners Support Network.[40] More darkly, like the loose religious sect that they essentially are, they also strongly shun former confederates who they see as having betrayed them and their ideals.[41]

Permaculture and *refugia*

More settled and literally down to earth than militant environmentalism "for the wild" is another approach that nonetheless shares its key Gaian premises. Similar in their rejection of anthropocentrism in favor of some manner of allegedly restorative holism, certain of these neo-pagan formations show interesting signs of vibrancy and an ability to fuse with modern science.

A compelling example arising from the larger and by-now mainstream organic farming movement is the practice of "permaculture," a term first coined and articulated by Australian ecologists Bill Mollison and David Holmgren in the 1970s. Permaculture has steadily gained adherents since. Mollison describes permaculture as "a philosophy of working with, rather than against nature; of protracted and thoughtful observation rather than protracted and thoughtless labour; and of looking at plants and animals in all their functions, rather than treating any area as a single product system."[42] It is not merely a philosophy, however. Permaculturalists are devoted to investigating and sharing techniques for agricultural (including forest) management with a dedication to trial-and-error scientific empiricism and a practical zeal for "what works." Yet it is also very high concept in that the notion of juxtaposing natural elements to achieve certain effects in order to create virtuous cycles in a garden, for example, is understood to run counter to the mainstream agricultural tendency toward industrial farming and mono-cropping and the prevailing scorched earth approaches to ecosystem-destroying large-scale use of herbicides and pesticides. It is very close to a way of life, a comprehensive conception in the above senses, because it orients one's way of thinking toward the positioning of natural elements and the incorporation of pre-industrial techniques.

If I have a pest problem with one of my garden vegetables, for example, I think in terms of what I can plant amidst or adjacent to the affected vegetable to control the problem rather than purchasing a chemical spray in order to eradicate everything living around the plant I want to protect. It could be as simple as planting horseradish at the base of a cherry tree to eliminate the need for weeding and to promote the growth of the fruit.[43] Or, from my own personal experience with horses, harrowing their manure over their grazing pastures, in order to fertilize those very same pastures, creates a little cycle of grass-equine virtue that is good for everyone involved. This is the permaculture magic of simple distribution: as anyone with dogs knows, poop clumped together kills the grass underneath, whereas that same poop spread out fertilizes and makes the grass greener— and with the horses the cycle is complete because they then eat the grass that generates the manure, etc. Or there are more complicated examples that require longer-term planning like setting up a forest garden where large trees provide a canopy to soak up the sun to create optimal shady conditions for berry bushes underneath, while vines crawl up the trees as ladders toward the sun (to mention just a few possible elements.) There are also scalable projects in urban permaculture involving whole villages and even cities where, for example, rooftop and balcony gardens can be nourished with gravity-fed gray water filtration and reclamation systems.

It is in this sense a less myopic and more holistic and relational mentality regarding both agricultural production and also simply modes of human and nonhuman animal sustenance and habitation generally. Importantly, though permaculture often looks to pre-industrial practices for practical techniques, it is not anti-modern or anti-science. Not in the least. Permaculturalists are typically extremely open to new ideas and efforts that might look crazy but have a sound scientific basis—as long as they *work*.

Another fascinating example of a salutary Gaian marriage of scientific and archaic elements is found in the description of "sacred groves" in India and elsewhere as found in the work of ecological anthropologist Debal Deb.[44] A nearly universal phenomenon across archaic cultures—Asia, Europe, Africa, America, Oceania—sacred groves were geographical spots, often in forests (but also in other places), that were regarded by local traditional cultures—"ecosystem peoples" in Deb's terminology—as untouchably sacred and so commonly zealously protected by special religious taboos and rules regarding the harvesting of grove plants and animals. Deb describes the basic idea: "Village communities used to consider distinct patches of vegetation, tanks and ponds as sacred, where any and all acts potentially desecrating or damaging to the habitat were customarily forbidden. Relics of thousands of these sacred groves still exist in most parts of the Indian subcontinent, unless expunged by 'development' through agricultural extension, industrialization, or urban sprawl."[45] These sacred groves became treasure houses of biodiversity of the local plants and animals, *refugia*, as Deb terms them.[46] It is thought that many plants and animals of today have survived because, at least at certain times, they were preserved and protected within the relative safety of a sacred grove. It is a particularly fascinating phenomenon because, in a way like permaculture, it can be understood as a fusion of archaic and traditional practices and beliefs—many no doubt regarded by the modern mind as "irrational" (e.g., like preferring starvation to killing a deer in a sacred grove)—yet nonetheless possessing these undeniable salutary concrete consequences, in this case regarding the preservation of biodiversity.

One of Deb's most important points is that we need to eliminate a modernist prejudice against "local people's knowledge and concern" as inferior when as a rule "species and ecosystems have co-evolved with humans" and that the ecosystem peoples were relatively non-disruptive and superior managers of local resources.[47] I cannot resist suggesting that this is an example of *conservatism* in the truest and most beneficial sense. It is also a model for how archaic ritual practices and beliefs might be reconstituted for modern mentalities. A taboo surrounding a modern *refugium* might not need to be reinforced by spells and demons but rather by a commitment concerning the delicateness and preciousness of the lives it is containing within its confines, the salvific powers of the grove therefore being appreciated in both "scientific" and "sacred" modes of understanding.

However promising, devotees of any of the movements or practices within this category—from neo-paganism and Wiccan to remnants of indigenous belief systems and contemporary permaculturalists—are quite small as a percentage of the population. It is hard to say but taken all together there are probably no more than a few million.[48] In any event, they are a decided minority. Despite their long history—even in the United States[49]—unlike a conventional religion like Christianity, then, the neo-pagan et al. nature-oriented devotees simply do not have the cultural presence and collective resonance necessary at present to provide the undergirding for a mass social movement. Yet from within this category could emerge a dark horse belief system, especially if some of the apocalyptic forecasts regarding climate change are realized. One must also remember that every major world religion was once held only by a small minority and regarded as eccentric. I would suggest viewing this category as small but very much alive, in the manner of a time capsule that is available to be opened in times of great planetary stress, in that sense, appropriately enough, almost as a *cultural* refugium within which certain potentially invaluable ideas and practices are being preserved at a small scale as a kind of hedge. Under the right (or horribly wrong) atmospheric conditions, there could well be a salvific power—permaculture, say, if petroleum grows scarce and thus also fertilizer and the efficiencies of mechanized farming—standing in reserve to go viral as needed among the remnants of whatever post-catastrophic scenario obtains.

Epilogue

*I like it when a flower or a little tuft of grass grows through a crack
in the concrete. It's so fuckin' heroic.*
George Carlin (1997)[1]

A brief summary is in order as this book presents a complicated
argument. Human understanding is always a bidirectional
movement between part and whole where remaining at one pole is
to lose sight. At its best but hardly exclusively, philosophy can help
instigate moves in either direction. What is needed at this moment
is greater attention to big picture issues that are being under-
discussed in, specifically, the area of the moral underpinnings
of political motivation. This book will be successful if even in
disagreement it helps move the reader's mind that way. It is all
part of what adrienne maree brown calls "emergence," which is
a strategy of engaging in simple interactions in order to facilitate
larger patterns—in this case a particular kind of conversation—on
the basis of a conviction that "what you pay attention to grows."[2]
The goal of this book is thus to participate with others in altering
our current attention economy.

Our setting is the moral aftermath of neoliberal capitalism
and the sense of existential drift occasioned when we no longer
feel ourselves needed in the jobs that had previously anchored
us into society. This sense of being disconnected and purposeless
manifests itself in many ways but especially in a proliferation of
the basic identity categories through which we know ourselves.
These non-vocational identities are initially seized upon with
a special zeal because they help fill the psychological void left
behind by the departure of stable occupational identities. This
is alarming and exhilarating and sometimes just plain amusing
(if it is not possible to keep a sense of humor then all is lost)
and it manifests itself on all parts of the political spectrum. As

they become apparent in political agendas, in their ensemble the proliferated identities start to churn the social world and individuals consequently experience an existential vertigo where what they once took as solid melts all around them. We feel that we are floating unanchored and this causes us to cling ever more vigorously to our identity flotsam during the economic and cultural storms or else we lose hope and sink.

But the identity liquidity can only last so long and go so far; we are not wired to sustain this churning indefinitely. By "wired" I do not mean some kind of vulgar biological determinism. I am asserting only the barest of philosophical anthropologies with three baseline components: 1) we are narrative creatures who *seek meaning* in our lives; 2) we *exist in time* and are mortal; and 3) we are *aware of our mortality* (in Heideggerian language being such as us have the ontic characteristic of being ontological).[3] Following Bauman, these ingredients add up to the contention that liquid modernity is not sustainable indefinitely; we are likely in a sort of interregnum period prior to some new and as-yet-unforeseen dispensation of identity categories. Stated dramatically, we are currently in the process of liquidizing ourselves—our trans-human moment—as a prelude to our being poured into new identity molds. A more enthusiastic transhumanist might suggest that because we are liquidizing now that therefore we can liquidize forever. I think this is ahistorical and assumes an overly plastic conception of the self that is actually a crypto-theological residue of belief in a transcendent soul. The current ideological stakes are therefore very high, higher than they may seem. For at present we are already undergoing—knowingly or not—battles over the shape of those awaiting identity molds and the metanarratives that will determine them.

These "imagination battles" are being waged across all cultural spheres: the tectonic plate shifting of the economic base of society touches them all—my analysis remains "Marxist" in this general sense. Prominent among these spheres are education, politics and art, all broadly construed. Given my own background and expertise, my analysis focuses on the first two while drawing selected examples from the latter (among others). Education is often considered the primary area in which a culture deliberately attempts to construct and inculcate its authorized identities. When it does so in an intensely aware manner it conducts a *paideia*, where it seeks to reproduce an acknowledged cultural ideal. Neoliberal nihilism has however rendered this impossible. There are vestiges but at this point the only guiding principle in MEDC education systems is a very crude economism, the American and English systems being perhaps the world's worst. As I argued in this book's predecessor, the K-12 system is so far gone along these lines as to be un-reformable.

If anyone were tempted to posit the university system as a saving grace, the situation there is even worse than in primary education; it is a system that is even more hopelessly driven by bottom-line considerations—even amidst its delusions of grandeur. With its full embrace of market myopia, our higher education system unknowingly commits itself to the wholesale commodification of learning (qua future earning power) as well as the instrumentalization of students themselves (qua customers and business resources). Everyone is reduced to market value and when automation drives more and more toward zero value, we have more and more useless people—people who have been *conceived* as useless. Contrary to their public relations rhetoric, then, universities play a key role in the manufacture of uselessness, especially among those unfortunates who have accepted that their highest value lies in their economic utility (that is, most of us). Contrary to the official rhetoric, it turns out what we useless people need is a *real* educational process that will wean us off the nihilism and onto a sustaining worldview that speaks to our deepest selves; we need, as they say, *to get a life*.

So might that process of re-grounding ourselves take place

through politics, which especially in recent centuries has functioned as a major shaper of identities and metanarratives? The customary poles within which contemporary politics plays out are between left and right, the former consistently pressing claims of equality and inclusion and the latter consisting of an uneasy jumble of free market economics and the maintenance of traditional cultural attitudes. In short, though both sides style themselves morally superior, the left's egalitarianism renders it unable to ground its commitments in any particular worldview, for this would not be inclusive of other world views. At the same time, the right refuses to acknowledge how laissez-faire economics undermines the cultural stasis (or atavism) it claims to defend. In different ways bereft of their traditional worldview supporting systems (viz., widely-shared religious moralities), all parts of our thinned-out political spectrum are susceptible to the rising identity tide of liquid modernity. For its part, in the face of this challenge, the right has become ideologically incoherent, a hodge-podge of market worship, faux traditionalism and simmering cultural resentments, as the 2016 Trump campaign exemplified.

Focusing on the left, what is observable is that a nihilistic and authoritarian identity politics in the corporate and academic worlds threatens to disfigure any recognizable commitment to egalitarianism. But without the egalitarianism, there is no longer any "left" left. Unknowingly following the logic of identity proliferation, identitarians break the world apart into mini-worldviews that represent a devolution into a serial authoritarianism, little careerist fortresses that cultural critic Mark Fisher derided as "the vampire castle," where progressive moral impulses are all-too-often channeled.[4] Absent properly universalizing countertrends such as social unionism or class politics (both alive but on life support), the construction of the vampire castle is an elaborate trap laid for the egalitarian left.

The difficult part is that what is needed to resuscitate

that project is by its very nature *outside* of the left and in fact outside of politics altogether: a grounding worldview capable of *basing* the left's egalitarian commitments. (One of the last of these morally-based leftward movements on the American scene was the civil rights movement of the 1950s and 60s, which was deeply rooted in the African-American Church.) Emotive criticisms of everyone's moral shortcomings will not accomplish this because those criticisms need to be *coming from somewhere* in order to be durably compelling. This need exposes the historical left's greatest weakness: minus some utopian experiments, the left—and particularly the Marxist left—has consistently been strong on critique but woefully lacking in the development of any positive successor vision. The refusal to make explicit its own normative underpinnings is both cause and symptom of this inability to construct such a vision. If it is not to abandon the field in favor of the political right's inevitable regression into blood and soil, the left must take an ideological step back and consider what it would take to advance a proper worldview in which it could convincingly base its moral claims. Because it offers *some* psychological sustenance, a "bad worldview" will defeat "no worldview" every time. As journalist George Monbiot emphasizes in his compelling book *Out of the Wreckage*, "It is not enough to challenge an old narrative, however outdated and discredited it may be. Change happens only when you replace it with another. When we develop the right story and learn how to tell it, it will infect the minds of people across the political spectrum."[5] This book's goal is thus to move the debate out of the arena of "the critique of capitalism and neoliberalism"— as important as accomplishing that has been—and, along with Monbiot, into the more constructive and hence riskier territory of telling "the right story."

So I describe what I see as the available options for doing just that. It is really just basic rationality that, after all, involves merely the adequation of means to ends. There is no general defense of rationality itself because on the above definition it is but a morally neutral relational concept. Moral defensibility depends independently upon the ends, analogous to how the premises of an argument must be established independently from that same argument's validity. At the very least, for one's politics to be rational, one must at least *have* a worldview — *some* worldview — within which one's normative claims make sense. Rationality thus has an inevitable social dimension having to do with one's ability to convince someone of one's proposed course of action. In this sense a serial killer with a developed rationale for serial killing (if such a thing were possible) would, unfortunately, be more rational than a person going around making impassioned moral claims with no sense whatever of what those claims might be based on. My simple plea then is that the left should attempt to become more rational, that is, it should develop a more considered view of the *ends* of its activity, in order for it to be more convincing to others. Otherwise the serial killer wins. Yet paradoxically in order to become more rational, what is needed is a protracted round of *irrationality*; if rationality is adequation of means to ends, then a necessary condition for rationality is the *having* of ends. But ends, like premises, are to be established outside of the adequation process — they have to appear as given assumptions in the overall argument. To be properly political, then, one must voyage outside of politics and then import back into politics one's vision of the world. A holistic worldview must be articulated, defended and *pursued* via the political rather than be *constituted* by it.

After going through the options I attempt to put my money where my mouth is and propose one that seems especially compelling: the multi-level Gaian approach, various elements of which are sketched in the previous chapter. As I hope to have indicated, it would be absurd to posit an exclusive claim to truth for any one perspective on the largest questions of human existence. This premise of pluralism strengthens the appeal of the Gaian framework because it admits of many different portals of entry, including those that are amenable to what I argue is the political left's need for a Nietzschean extrapolation of egalitarianism, where the substratum comprising that to which moral concern adheres is extended. As for exactly what politics falls out of all this, I can only reiterate my caveat that there can be no guarantees. I will also echo what Timothy Morton says about his own fascinating parallel investigation that "Yes, it's possible to include nonhuman beings in Marxist theory—but you're not going to like it!"[6] Well, maybe you will maybe you won't. What remains is to explore together further contours of the Gaian worldview to gain a clearer sense of its antecedents and the many directions it might lead. We have, in other words, a worldview to win.

Notes

i.   adrienne maree brown, *Emergent Strategy: Shaping Change, Changing Worlds* (AK Press, 2017), 18.

Preface

1.   Mark Fisher, "Exiting the Vampire Castle," *The North Star*, November 22, 2013, http://www.thenorthstar.info/2013/11/22/exiting-the-vampire-castle/.

2.   This is a theme throughout Dewey's work, e.g., John Dewey, *Liberalism and Social Action* in *The Later Works of John Dewey, Volume 11, 1925–1953* (Southern Illinois University Press, 2008).

3.   Amartya Sen, *Poverty and Famines: An Essay on Entitlement and Deprivation* (Oxford University Press, 1983).

4.   Walter Scheidel and Steven J. Friesen, "The Size of the Economy and the Distribution of Income in the Roman Empire." *The Journal of Roman Studies* 99 (2009): 61-91. "Scheidel and Friesen estimate that the top 1 percent of Roman society controlled 16 percent of the wealth, less than half of what America's top 1 percent control." See Central Intelligence Agency, "The World Factbook: GINI Measures." https://www.cia.gov/library/publications/the-world-factbook/fields/2172.html.

5.   Thomas Piketty, *Capital in the Twenty-First Century* (Harvard University Press, 2017), 663-666.

6.   "Positionality" here is meant in the sense of "positional good," in economics lingo a good whose value is proportional with its scarcity across a population. e.g., my college degree is worth less if everybody around me has one. The phrase was originated by Fred Hirsch, The Social Limits to Growth (Routledge & Kegan Paul, 1977).

7.   Kurt Vonnegut, *Player Piano* (New York: The Dial Press, 1999).

8.   Alexis C. Madrigal, "Silicon Valley's Big Three vs. Detroit's Golden-Age Big Three," *The Atlantic* (May 24, 2017), https://

www.theatlantic.com/technology/archive/2017/05/silicon-valley-big-three/527838/.

9. "In 2016, Facebook generated $600,000 of net income *per employee.*" *Op. cit.*

10. David Harvey, *Marx, Capital and the Madness of Economic Reason* (Oxford University Press, 2017).

11. Simon Mohun, "Class Structure and the US Personal Income Distribution, 1918–2012," *Metroeconomica*, vol. 67, no. 2 (May 2016): 334-363.

12. United States Holocaust Memorial Museum, "Nazi Persecution of the Disabled: Murder of the 'Unfit' (online exhibition)," retrieved April 2018, https://www.ushmm.org/information/exhibitions/online-exhibitions/special-focus/nazi-persecution-of-the-disabled.

13. Riley Woodford, "Lemming Suicide Myth: Disney Film Faked Bogus Behavior," *Alaska Fish & Wildlife News* (September 2003), http://www.adfg.alaska.gov/index.cfm?adfg=wildlifenews.view_article&articles_id=56.

14. Yair Rosenberg, "'Jews will not replace us': Why white supremacists go after Jews," *The Atlantic* (August 14, 2017); and Sarah Wildman, "'You will not replace us': a French philosopher explains the Charlottesville chant," *Vox* (August 15, 2017), https://www.vox.com/world/2017/8/15/16141456/renaud-camus-the-great-replacement-you-will-not-replace-us-charlottesville-white.

15. Andrew Yang, *The War on Normal People: The Truth About America's Disappearing Jobs and Why Universal Basic Income Is Our Future* (Hachette Books, 2018).

16. Joan Robinson, *Economic Philosophy* (Routledge, 2006), 45.

17. John Marsh, *Class Dismissed: Why We Cannot Teach or Learn Our Way Out of Inequality* (Monthly Review Press, 2011).

18. Plato, *Protagoras* 360d; *Phaedo* 84a; *Laches* 199c-e (among other places).

19. David Brond, "New Brand Showcases the Best of UD,"

University of Delaware Messenger vol. 18, no. 2, retrieved April 2018, http://www1.udel.edu/udmessenger/vol18no2/stories/feature_dare-to-be-first.html.

20. See Bryan Caplan, *The Case against Education: Why the Education System Is a Waste of Time and Money* (Princeton University Press, 2018); Josh Mitchell and Douglas Belkin, "Americans Losing Faith in College Degrees, Poll Finds," *The Wall Street Journal* (September 7, 2017), https://www.wsj.com/article_email/americans-losing-faith-in-college-degrees-poll-finds-1504776601-lMyQjAxMTI3NzAyODIwMzg2Wj/.

21. Jurgen Habermas, *Legitimation Crisis* (Beacon Press, 1975).

22. Martin Luther King Jr, "Beyond Vietnam: A Time to Break Silence," (New York, NY: April 4, 1967), http://inside. sfuhs.org/dept/history/US_History_reader/Chapter14/MLKriverside.htm.

23. Josh Hafner, "Trump Loves the Poorly Educated — And They Love Him," *USA Today* (February 24, 2016), https://www. usatoday.com/story/news/politics/onpolitics/2016/02/24/donald-trump-nevada-poorly-educated/80860078/.

24. *Idiocracy*, directed by Mike Judge (2011: 20th Century Fox). DVD.

25. G.W.F. Hegel, *Early Theological Writings* (University of Pennsylvania Press, 1971), 184-185.

26. John Rawls, *Political Liberalism* (Columbia University Press, 2005), 13. For a summary of this concept in the education context, see Judith Suissa, "How Comprehensive is Your Conception of the Good? Liberal Parents, Difference and the Common School," *Educational Theory*, vol. 60, no. 5 (October 2010): 587-600.

27. Aristotle, *Nicomachean Ethics*, Book X, available at *The Internet Classics Archive*, http://classics.mit.edu/Aristotle/nicomachaen.html.

28. Zygmunt Bauman, *Liquid Modernity* (Polity, 2000).

29. Matthew Charles, "*Gemeinspruch*: On Transdisciplinarity

in Education Theory," *Radical Philosophy* 183 (January/ February 2014): 61.

30. Kant, *Critique of Practical Reason* (Hackett Publishing Co., 2002), 5:161. 33–36.

Introduction

1. Yuval Noah Harari, "The Meaning of Life in a World Without Work," *The Guardian* (May 8, 2017), https://www.theguardian.com/technology/2017/may/08/virtual-reality-religion-robots-sapiens-book.

2. Pierre-Joseph Proudhon, *What is Property?* Available at *libcom.org*, https://libcom.org/files/Proudhon%20-%20 What%20is%20Property.pdf.

3. Marx, *Capital: Volume 1*, Chapter 10, Section 1, available at *Marxists Internet Archive*, https://www.marxists.org/archive/ marx/works/1867-c1/ch10.htm.

4. Bauman, *Liquid Modernity*, 139.

5. "Dead zone" is a phrase from marine ecology describing ocean areas that have become uninhabitable, usually from some form of pollution. Some have extended it metaphorically to describe areas where people have been "sacrificed" to extremely suboptimal economic conditions. For example, see David Perlman, "Scientists alarmed by ocean dead-zone growth," *SFGate* (August 15, 2008), http://www.sfgate.com/green/article/Scientists-alarmed-by-ocean-dead-zone-growth-3200041.php and Henry A. Giroux, "In the Dead Zone of Capitalism: Lessons on the Violence of Inequality from Chicago," *Truthout* (June 4, 2003), http://www.truth-out.org/opinion/item/16738-in-the-dead-zone-of-capitalism-lessons-on-the-violence-of-inequality-from-chicago.

6. Joshua Greene, *Moral Tribes: Emotion, Reason, and the Gap Between Us and Them* (Penguin Books, 2014) and Jonathan Haidt, *The Righteous Mind: Why Good People Are Divided by*

Politics and Religion (Vintage Books, 2013).

7. See for example Nicolas Boyon and Julia Clark, *The New Tribalism: Clashing Views on Who Is a Real American* (IPSOS Public Affairs, 2017), https://www.ipsos.com/en-us/news-polls/new-tribalism-2017-10; Carol S. Pearson, "Finding Yourself in the New Tribalism," *Psychology Today* (March 23, 2017), https://www.psychologytoday.com/blog/the-hero-within/201703/finding-yourself-in-the-new-tribalism. Though perhaps "the new tribalism" is not so new: I.H.T. Editorial, "May 21, 1968: The New Tribalism," *New York Times* (May 21, 1968), http://www.nytimes.com/2013/10/14/opinion/14iht-op1968may21.html.

8. Friedrich Engels, *Socialism: Utopian and Scientific*, available at *Marxists Internet Archive*, https://www.marxists.org/archive/marx/works/1880/soc-utop/index.htm.

9. I explore this at length in David J. Blacker, *Democratic Education Stretched Thin: How Complexity Challenges a Liberal Ideal* (SUNY Press, 2007).

10. For overviews, see Sarah Azaransky, *This Worldwide Struggle: Religion and the International Roots of the Civil Rights Movement* (Oxford University Press, 2017) and Gary Dorrien, *Breaking White Supremacy: Martin Luther King Jr. and the Black Social Gospel* (Yale University Press, 2018).

11. Martin Luther King Jr, "I've Been to the Mountaintop," speech delivered April 3, 1968, Memphis, TN, available at http://www.americanrhetoric.com/speeches/mlkivebeentothemountaintop.htm.

12. Martin Luther King, Jr, "Address at the Freedom Rally in Cobo Hall," speech delivered June 23, 1963, Detroit, MI. Available at https://kinginstitute.stanford.edu/king-papers/documents/address-freedom-rally-cobo-hall.

Chapter 1

1. Eric Hobsbawm, "The Cult of Identity Politics," *New Left Review* 217 (1998): 40.

2. Immanuel Kant, *Grounding for the Metaphysics of Morals* (Hackett Publishing Co., 1993).

3. Karl Marx, *Theses on Feuerbach*, available at *Marxists Internet Archive*, https://www.marxists.org/archive/marx/works/1845/theses/theses.htm.

4. Martin Ford, (New York: Basic Books, 2016); Tim Dunlop, (New South Publishing, 2016); Jerry Kaplan (Yale University Press, 2016); Larry Elliott, "Robots will take our jobs. We'd better plan now, before it's too late," *The Guardian* (February 1, 2018), https://www.theguardian.com/commentisfree/2018/feb/01/robots-take-our-jobs-amazon-go-seattle.

5. Yang, *The War on Normal People*, 1.

6. Zygmunt Bauman, *Wasted Lives: Modernity and its Outcasts* (Polity, 2003).

7. Friedrich Nietzsche, *The Twilight of the Idols and the Anti-Christ* (Penguin Books, 1990).

8. See David J. Blacker, *The Falling Rate of Learning and the Neoliberal Endgame* (Zero Books, 2013).

9. Saskia Sassen, *Expulsions: Brutality and Complexity in the Global Economy* (Harvard University Press, 2014).

10. Matthew 8:12, 22:13, and 25:30.

11. David Hume, *A Treatise of Human Nature* (The Clarendon Press, 1973), 577.

12. Stephen Darwall, *The Second-Person Standpoint: Morality, Respect, and Accountability* (Harvard University Press, 2009).

13. Immanuel Kant, "On the Original Predisposition to Good in Human Nature," in *Religion Within the Bounds of Bare Reason* (Hackett Publishing Co., 2009), 27.

14. Jean-Paul Sartre, *Existentialism is a Humanism*, available at *Marxists Internet Archive*, https://www.marxists.org/reference/archive/sartre/works/exist/sartre.htm.

15. See John Stuart Mill, *Utilitarianism* (1863), Chapter 5, available at https://www.utilitarianism.com/mill5.htm and Peter Singer, "Why Care about Equality?" *Cato Unbound* (March 7, 2006), available at https://www.cato-unbound. org/2006/03/07/peter-singer/why-care-about-equality-response-schmidtz.

16. Kant, *Grounding for the Metaphysics of Morals*, 4:422-423.

17. Kant, *The Metaphysics of Morals* (Cambridge University Press, 1996), 6:392.

18. Declaration of Independence (US, 1776).

19. "If I am captured I will continue to resist by all means available. I will make every effort to escape and to aid others to escape. I will accept neither parole nor special favors from the enemy." (*Code of Conduct for Members of the United States Armed Forces*, Article III, Executive Order 10631) (1955). The Geneva Convention also recognizes this duty and places restrictions on the treatment of recaptured escapees (Articles 42, 91-93).

20. G.W.F. Hegel, *Phenomenology of Spirit* (Oxford University Press, 1977), 111f. Among his long list of cultural universals, anthropologist Donald Brown includes relevant items like "reciprocity," "recognition," the desire for a "positive self-image," "shame," etc. See Donald E. Brown, *Human Universals* (McGraw-Hill, 1991) and "Human universals, human nature & human culture," *Daedalus*, vol 22, no. 4 (Fall 2004): 47-54.

21. Michael Bond, "How extreme isolation warps the mind," *BBC Future* (March 14, 2014), http://www.bbc.com/future/story/20140514-how-extreme-isolation-warps-minds; Judith Shulevitz, "The Lethality of Loneliness," *New Republic* (May 13, 2013), https://newrepublic.com/article/113176/science-loneliness-how-isolation-can-kill-you.

22. Aristotle *Politics*, Book I, 1253a, available at *The Internet Classics Archive*, http://classics.mit.edu/Aristotle/politics.

html.

23. See Donald Brown, op. cit.

24. Rawls, *A Theory of Justice* (Harvard University Press, 1971), 440.

25. Ibid.

26. One may find voluminous sayings in this vein, see *Goodreads*, http://www.goodreads.com/quotes/tag/self-respect.

27. Rawls, *Theory of Justice*, 441.

28. Thomas E. Hill, Jr, "Stability, a Sense of Justice, and Self-respect," in Jon Mandle and David E. Reidy, eds, *A Companion to Rawls* (New York: John Wiley & Sons, 2014), 206.

29. Rebecca Riffkin, "In US, 55% of Workers Get Sense of Identity From Their Job," *Gallup* (August 22, 2014), http://www.gallup.com/poll/175400/workers-sense-identity-job.aspx.

30. Ibid.

31. "In all world regions, the relative risk of suicide associated with unemployment was elevated by about 20–30% during the study period." Carlos Nordt et al., "Modelling suicide and unemployment: a longitudinal analysis covering 63 countries, 2000–11," *The Lancet Psychiatry*, vol. 2, no. 3 (March 2015): 239–245; Rutgers University researchers found that "external economic factors were present in 37.5 percent of all suicides in 2010." See Katherine Hempstead and Julie Phillips, "Rising Suicide Among Adults Aged 40–64 Years: The Role of Jobs and Financial Circumstances," *American Journal of Preventative Medicine*, vol. 48, no. 5 (May 2015): 491–500.

32. See Hempstead and Phillips, op. cit., who also note, "relative to other age groups, a larger and increasing proportion of middle-aged suicides have circumstances associated with job, financial, or legal distress." See also Daniel Schwartz, "Suicide rates are highest for men in their 50s and we're not sure why," *CBC News* (April 27, 2016), http://www.cbc.ca/

news/health/suicide-men-50s-causes-1.3263412.

33. Bauman, *Liquid Modernity*, 161.

34. David Callahan, "The Most Bogus Unemployment Number: Discouraged Workers," *Demos* (September 6, 2014), http://www.demos.org/blog/9/6/13/most-bogus-unemployment-number-discouraged-workers.

35. For example, left-wing polemicist Laurie Penny claims on first-hand knowledge that her frenemy (for lack of a better term), the alt-right superstar Milo Yiannopoulos, actually believes little or nothing of the ideology he professes in public. "I'm With the Banned," Laurie Penny, *medium.com* (July 21, 2016), https://medium.com/welcome-to-the-scream-room/im-with-the-banned-8d1b6e0b2932.

36. David Harvey, *Marx, Capital and the Madness of Economic Reason*, 196-197.

37. "On the Phenomenon of Bullshit Jobs," *Strike! Magazine* (August 17, 2013), https://strikemag.org/bullshit-jobs/.

38. Ibid.

39. See "'The Office' Quotes," available at http://m.theofficequotes.com/?url=http%3A%2F%2Fwww.theofficequotes.com%2Fseason-4%2Fjob-fair&utm_referrer=#2660.

40. Bauman, *Liquid Modernity*, 148.

41. Ibid.

42. Rick Warren, *The Purpose Driven Life: What on Earth Am I Here For?* (Zondervan, 2012).

43. Bauman, *Liquid Modernity*, 139-140.

44. By "neo-artisanal" I mean occupations such as organic gardening or permaculture or other production processes that arise out of committed ideological niches and/or deal with emerging materials that meet novel consumer demands, e.g., certain forms of digital art or web design or, perhaps, catering to a market based on nostalgia for a "lost art" where a priority is given to supporting and recovering

an "authentic" production process. Some of this sensibility is chronicled in David Sax, *The Revenge of the Analog: Real Things and Why They Matter* (Public Affairs, 2016).

45. Friedrich Nietzsche, *The Gay Science* (Vintage, 1974), 125.

46. *Communist Manifesto*. Available at https://www.marxists. org/archive/marx/works/1848/communist-manifesto/ch01. htm.

47. Ha Jin *War Trash* (Vintage, 2005).

48. Nicholson Baker, *Human Smoke: The Beginnings of World War II, the End of Civilization* (Simon & Schuster, 2009).

49. Justin Scholes and Jon Ostenson, "Understanding the Appeal of Dystopian Young Adult Fiction," *The ALAN Review*, vol. 40, no. 2 (Winter 2013): 11-20; Lauren Sarner, "Dystopian fiction, and its appeal: Why do apocalyptic portrayals of existence dominate teen shelves?," *New York Daily News* (June 28, 2013), http://www.nydailynews.com/blogs/pageviews/ dystopian-fiction-appeal-apocalyptic-portrayals-existence-dominate-teen-shelves-blog-entry-1.1640750; Christopher Schmidt, "Why are Dystopian Films on the Rise Again?" *JSTOR Daily* (November 19, 2014), https://daily.jstor.org/ why-are-dystopian-films-on-the-rise-again/.

50. Walter R. Fisher, *Human Communication as Narration: Toward a Philosophy of Reason, Value, and Action* (University of South Carolina Press, 1989); H. Porter Abbott, *The Cambridge Introduction to Narrative*, 2nd ed. (Cambridge University Press, 2008).

51. Roland Barthes, *An Introduction to the Structural Analysis of Narrative* (University of Birmingham, 1966) as quoted in Lionel Duisit, "On Narrative and Narratives," *New Literary History*, vol. 6, no. 2 (Winter, 1975): 237.

52. Rawls, *Political Liberalism*, 144.

53. Kimberle Crenshaw, "Demarginalizing the Intersection of Race and Sex: A Black Feminist Critique of Antidiscrimination Doctrine, Feminist Theory and Antiracist Politics," *University*

of *Chicago Legal Forum*, vol. 140 (1989): 139-167.

54. Jean-Paul Sartre and Bernard Frechtman, *Saint Genet: Actor and Martyr* (University of Minnesota Press, 2012), 49.

55. "men never have been and never will be able to undo or even control reliably any of the processes they start through action," Hannah Arendt, *The Human Condition* (University of Chicago Press, 1958), 190.

56. For understandable reasons this is a popular online topic, e.g., Barry J. Jacobs, "Caregivers: Living With Guilt," *AARP* (2017), https://www.aarp.org/caregiving/life-balance/info-2017/living-with-guilt-bjj.html (retrieved April 2018).

57. Bauman, *Liquid Modernity*, 38.

58. Keisha Mawer, "Tumblr's Strange Trend Of Creating Minority Identities," *Thought Catalogue* (September 3, 2014), http://thoughtcatalog.com/keisha-mawer/2014/09/tumblrs-strange-trend-of-creating-minority-identities/

59. See Sean Dunne's documentary, *American Juggalo*, directed by Sean Dunne (2011), https://www.imdb.com/title/tt2062478/; and Jon Ronson, "Insane Clown Posse: And God created controversy," *The Guardian* (October 8, 2010), https://www.theguardian.com/music/2010/oct/09/insane-clown-posse-christians-god

60. "Gender Master List," Genderfluid Support, accessed April 2018, http://genderfluidsupport.tumblr.com/gender.

61. "Sexual Orientations Masterlist," The Post Helpers, accessed April 2018, http://poshhelpers.tumblr.com/post/41062616375/sexual-orientations-masterlist.

62. *Fetlife*, retrieved April 2018, available at fetlife.com.

63. Anna Smith, "Fifty Shades of Grey: what BDSM enthusiasts think," *The Guardian* (February 15, 2015), https://www.theguardian.com/film/2015/feb/15/fifty-shades-of-grey-bdsm-enthusiasts; Stephanie Marcus, "'Fifty Shades Of Grey' Isn't A Movie About BDSM, And That's A Problem," (February 16, 2015), http://www.huffingtonpost.

com/2015/02/16/fifty-shades-of-grey-isnt-bdsm_n_6684808. html; Dylan Love, "This BDSM community is furious about 'Fifty Shades of Grey'" (February 26, 2015), http://www. dailydot.com/irl/fifty-shades-grey-bad-bdsm/; Cheyenne Picardo and Phoebe Reilly, "Whip Smart: Real-Life Dominatrix Takes on 'Fifty Shades of Grey'" (February 13, 2015), http://www.rollingstone.com/movies/features/real-life-dominatrix-takes-on-fifty-shades-of-grey-20150213.

64. Jean-Francois Lyotard, *The Postmodern Condition: A Report on Knowledge* (University of Minnesota Press, 1984), xxiv.

65. Robert Musil, *The Man Without Qualities* (Vintage, 1996).

66. For an overview see Joylon Mitchell, *Martyrdom: A Very Short Introduction* (Oxford University Press, 2013).

67. The Mahayana Buddhist monk who immolated himself in protest in Saigon (1963). Quang Duc's last words: "Before closing my eyes and moving towards the vision of the Buddha, I respectfully plead to President Ngo Dinh Diem to take a mind of compassion towards the people of the nation and implement religious equality to maintain the strength of the homeland eternally. I call the venerables, reverends, members of the sangha and the lay Buddhists to organize in solidarity to make sacrifices to protect Buddhism." His death produced iconic images from this period. See "The burning monk, 1963," *Rare Historical Photos* (June 23, 2015), http://rarehistoricalphotos.com/the-burning-monk-1963/.

68. Martin Luther King, Jr, "Speech at the Great March on Detroit," June 23, 1963. http://kingencyclopedia.stanford. edu/encyclopedia/documentsentry/doc_speech_at_the_great_march_on_detroit.1.html.

69. I have explored this theme at length in David J. Blacker, *Dying to Teach: The Educator's Search for Immortality* (Columbia University Teachers College Press, 1997).

Chapter 2

1. Werner Jaeger, *Paideia: The Ideals of Greek Culture: Volume I: Archaic Greece: The Mind of Athens* (Oxford University Press, 1986), xiv.

2. I defend this thesis in Blacker, *The Falling Rate of Learning and the Neoliberal Endgame.*

3. Bureau of Labor Statistics, "Labor Force Statistics from the Current Population Survey," (US Department of Labor), data extracted April 18, 2018, http://data.bls.gov/timeseries/ LNS11300000. Some argue that the numbers only look bad because of demographics, viz., baby boomer retirees.

4. Derek Thompson, *The Atlantic,* "What Are Young Non-Working Men Doing?" *The Atlantic* (July 25, 2016), http:// www.theatlantic.com/business/archive/2016/07/what-are-young-non-working-men-doing/492890/.

5. Tom Peters, "The Brand Called You," *Fast Company* (August 31, 1997), https://www.fastcompany.com/28905/brand-called-you.

6. "What else does this craving, and this helplessness, proclaim but that there was once in man a true happiness, of which all that now remains is the empty print and trace? This he tries in vain to fill with everything around him, seeking in things that are not there the help he cannot find in those that are, though none can help, since this infinite abyss can be filled only with an infinite and immutable object; in other words by God himself." Blaise Pascal, *Pensées* (Penguin, 1995), VII, 425.

7. Thomas Jefferson, *Notes on the State of Virginia* (1785), available at http://web.archive.org/web/20110221130550/ http://etext.lib.virginia.edu/etcbin/toccer-new2?id=JefVirg. sgm&images=images/modeng&data=/texts/english/ modeng/parsed&tag=public&part=all.

8. George Eliot, *Silas Marner*, available at Project Gutenberg, http://www.gutenberg.org/ebooks/550.

9. This has been a defining theme of First Amendment religion clauses jurisprudence in the US since the 1960s. It has involved the prohibition of school-sponsored prayer and devotional Bible reading in public schools (*Engel v. Vitale*, 370 US 421 (1962); *Schempp v. Abington Township*, 374 US 203 (1963); *Lemon v. Kurtzman*, 411 US 192 (1973)), concerns about the "coercive" effect of religious practices even in extracurricular activities (*Lee v. Weisman*, 505 US 577 (1992); *Santa Fe v. Doe*, 530 US 290 (2000)), and the thwarting of efforts by religious fundamentalists to ban the teaching of evolution (*Epperson v. Arkansas*, 393 US 97 (1968)) and to require "creationist" Bible narratives to counter it in science classes (*Edwards v. Aguillard*, 482 US 578 (1987)).

10. *Stone v. Graham*, 449 US 39 (1980) (posting the Ten Commandments in public school classrooms violates the Establishment Clause) and *Lautsi v. Italy*, App. No. 30814/06, 2011 Eur. Ct. H.R. (G.C.) (posting crucifixes in public classrooms violates state religious neutrality). Two years previously, the European Court had originally ruled for Italy on the basis of the legitimacy of preserving its "cultural heritage."

11. Werner Jaeger, *Paideia: The Ideals of Greek Culture*, 4.

12. Catherine Rampell, "A chilling study shows how hostile college students are toward free speech," *Washington Post* (September 18, 2017), https://www.washingtonpost.com/opinions/a-chilling-study-shows-how-hostile-college-students-are-toward-free-speech/2017/09/18/cbb1a234-9ca8-11e7-9083-fbfddf6804c2_story.html?utm_term=.553d2675c7f4.

13. University of Delaware, www.udel.edu.

14. John Dellacontrada, "UB ready to tell its story worldwide — with Buffalo at the heart of it," *UB Now* (April 12, 2016), http://www.buffalo.edu/ubreporter/stories/2016/04/brand-launch.html. For a brief overview of this trend, see Ellen

Wexler, "Your Future Starts Here. Or Here. Or Here," *Inside Higher Education* (May 2, 2016), https://www.insidehighered.com/news/2016/05/02/why-colleges%E2%80%99-brands-look-so-similar.

15. See Richard Arum, *Academically Adrift: Limited Learning on College Campuses* (Chicago: University of Chicago Press, 2011) and William Deresiewicz, *Excellent Sheep: The Miseducation of the American Elite and the Way to a Meaningful Life* (New York: Free Press, 2015).

16. Bryan Caplan, *The Case Against Education: Why the Education System is a Waste of Time and Money* (Princeton University Press, 2018), 13-15.

17. Hirsch, Fred. The Social Limits to Growth (Routledge & Kegan Paul, 1977).

18. This is a theme throughout Dewey's corpus, for example, *The Public and Its Problems* (Penn State Press, 2012), *Democracy and Education* (Simon & Brown, 2012), and *Liberalism and Social Action* (Prometheus Books, 1999).

19. Robert H. Frank and Philip J. Cook, *The Winner-Take-All Society: Why the Few at the Top Get So Much More Than the Rest of Us* (Penguin, 1996).

20. Nancy Schepler-Hughes, "The Crisis of the Public University," *Chronicle of Higher Education* (December 19, 2011), https://www.chronicle.com/article/The-Crisis-of-the-Public/130135.

21. Scott Jaschik, "Kentucky's Governor vs. French Literature," *Inside Higher Education* (February 1, 2016), https://www.insidehighered.com/quicktakes/2016/02/01/kentuckys-governor-vs-french-literature; Scott Jaschik, "Obama vs. Art History," *Inside Higher Education*, January 31, 2014, https://www.insidehighered.com/news/2014/01/31/obama-becomes-latest-politician-criticize-liberal-arts-discipline.

22. James Taranto, "Selling 'Diversity': The incentives behind higher education's bureaucratic bloat," *Wall Street Journal*

(December 2, 2011), https://www.wsj.com/articles/SB100014 240529702048267045770744831358O5796.

23. A "suspect class" is an American legal phrase denoting a "class of individuals that have been historically subject to discrimination." *Legal Information Institute*, Cornell University Law School, accessed April, 2018, https://www.law.cornell.edu/wex/suspect_classification.

24. Mark Dudzic and Adolph Reed, Jr, "The Crisis of Labor and the Left in the United States," *Socialist Register* (2015), 362.

25. Suzanne Collins, *The Hunger Games* (Scholastic Press, 2010).

26. "History," Neumann University, accessed April 2018, https://www.neumann.edu/about/history_mission.asp.

27. "Mission Statement," University of Notre Dame, accessed April 2018, https://www.nd.edu/about/mission-statement/.

28. "Mission and Vision Statement," Sacred Heart University, accessed April 2018, http://www.sacredheart.edu/faith service/officeofmissionandcatholicidentity/missionand values/missionstatement/.

29. "Vision & Mission," Regent University, https://www.regent.edu/about-regent/vision-mission/; and see Charlie Savage, "Scandal puts spotlight on Christian law school," *Boston Globe* (April 8, 2007), http://archive.boston.com/news/education/higher/articles/2007/04/08/scandal_puts_spotlight_on_christian_law_school/?page=1.

30. *Berea College v. Kentucky*, 211 US 45 (1908).

31. "Mission," Berea College, accessed April 2018, https://www.berea.edu/about/mission/.

32. "Mission and Values," Oberlin College, accessed April 2018, https://www.oberlin.edu/about-oberlin/mission-and-values. Oberlin tops at least one guide to "America's Most Liberal Colleges." For this and comparable examples, see Bestcolleges.com, "America's Most Liberal Colleges," available at http://www.bestcolleges.com/features/most-liberal-colleges/. The conservative list is also here, topped by

the nonsectarian Hillsdale College, available at http://www. bestcolleges.com/features/most-conservative-colleges/. Hillsdale's mission statement is interesting in that it is unusually polemical. It defends "the traditional liberal arts" and asserts that "The College values the merit of each unique individual, rather than succumbing to the dehumanizing, discriminatory trend of so-called 'social justice' and 'multicultural diversity,' which judges individuals not as individuals, but as members of a group and which pits one group against other competing groups in divisive power struggles." See "Mission," Hillsdale College, accessed April 2018, https://www.hillsdale.edu/about/mission/.

33. Allen Downey, "College Freshmen Are Less Religious Than Ever," *Scientific American* (May 25, 2017), https://blogs. scientificamerican.com/observations/college-freshmen-are-less-religious-than-ever/.

34. "Facts and Studies," *Council for American Private Education*, accessed April 2018, http://www.capenet.org/facts.html; "Number of college students in the US 1965-2026," *Statistica*, accessed April 2018, https://www.statista.com/ statistics/183995/us-college-enrollment-and-projections-in-public-and-private-institutions/.

35. John Dewey, *A Common Faith* (Yale University Press, 2013), 7.

36. Ibid., 87.

37. Daniel C. Dennett, "Cognitive Wheels: The Frame Problem of AI," in Christopher Hookway ed., *Minds, Machines and Evolution* (Cambridge University Press, 1984), 129-150.

38. Among many others, perhaps the foremost contemporary exponent of the view that universities have been taken over by "cultural Marxists" and, worse, "postmodernists," is University of Toronto psychology professor and Youtube star Jordan Peterson. See Jordan B. Peterson, "Postmodernism and Cultural Marxism," *Youtube* (July 6, 2017), accessed

April 2018, https://www.youtube.com/watch?v=s4c-jOdPT
N8&t=280s

39. Democratic Socialists of America, accessed April 2018,
   http://www.dsausa.org/.

40. Timothy Egan, "What if Steve Bannon Is Right?" *New
   York Times* (August 25, 2017), https://www.nytimes.
   com/2017/08/25/opinion/bannon-trump-polls-republican.
   html.

41. Roddy Forsyth, "Celtic vs. Rangers: The Old Firm
   Explained," *The Telegraph* (April 15, 2016), http://www.
   telegraph.co.uk/football/2016/04/15/celtic-vs-rangers-the-
   old-firm-explained/.

42. *Monty Python's Holy Grail*, directed by Terry Gilliam and
   Terry Jones, Michael White Productions, 1975.

43. *Aguirre, the Wrath of God*, directed by Werner Herzog, Werner
   Herzog Filmproduktion, 1972.

44. Putnam's extended metaphor is that there is as much
   bowling as ever but it is not being done as it once was in
   the more thickly affiliating setting of teams and bowling
   leagues. Robert D. Putnam, *Bowling Alone: The Collapse and
   Revival of American Community* (Touchstone Books, 2001).

45. Adolph Reed. Jr, "From Jenner to Dolezal: One Trans Good,
   the Other Not So Much," *Common Dreams* (June 15, 2015),
   https://www.commondreams.org/views/2015/06/15/jenner-
   dolezal-one-trans-good-other-not-so-much.

46. Richard Russo, *Straight Man* (Vintage, 1998).

47. Harry Frankfurt, *On Bullshit* (Princeton University Press,
   2005), 61.

48. Bhaskar Sunkara, "Let Them Eat Diversity: An interview
   with Walter Benn Michaels," *Jacobin* (January 1, 2011), https://
   www.jacobinmag.com/2011/01/let-them-eat-diversity.

Chapter 3

1. David Foster Wallace, *Brief Interviews with Hideous Men*

(Back Bay Books, 2000), 17.

2. Rebecca Reid, "Should 'K' for kinky be included under the LGBTQ+ umbrella?" *Metro.co.uk* (February 5, 2018), http://metro.co.uk/2018/02/05/k-kinky-included-lgbtq-umbrella-7282939/. The current acronym champion seems to be the Ontario Federation of Teachers, which conducts "inclusiveness training" utilizing "LGGBDTTTIQQAAPP" — and nonetheless fails to include the "K." See "Canadian elementary school teachers attend 'LGGBDTTTIQQAAPP' inclusiveness training session. Would you understand the title?" *Daily Mail* (November 28, 2017), http://www.dailymail.co.uk/news/article-5123013/Teachers-federation-conducts-LGGBDTTTIQQAAPP-training.html#ixzz5D8kZkKGz.

3. Daniel Cox, Rachel Lienesch, and Robert P. Jones, "Who Sees Discrimination? Attitudes on Sexual Orientation, Gender Identity, Race, and Immigration Status," *Public Religion Research Institute* (June 21, 2017), https://www.prri.org/research/americans-views-discrimination-immigrants-blacks-lgbt-sex-marriage-immigration-reform/.

4. Kevin Donovan and Alyshah Hasham, "Jian Ghomeshi lawsuit says CBC made 'moral judgment' about his sex life," *Toronto Star* (October 27, 2014). https://www.thestar.com/news/canada/2014/10/27/jian_ghomeshi_lawsuit_says_cbc_made_moral_judgment_about_his_sex_life.html. Matters seem to have grown murkier in this case as there are complicated allegations of non-consensuality in Ghomeshi's relationships which if true would make this not a matter of his BDSM sexual identity at all. See Brenda Crossman, "The Ghomeshi question: The law and consent," *Toronto Globe and Mail* (October 27, 2014, updated April 3, 2018), https://www.theglobeandmail.com/opinion/the-ghomeshi-question-the-law-and-consent/article21315629/.

5. Angela Nagle, *Kill All Normies: Online Culture Wars from 4Chan and Tumblr to Trump and the Alt-Right* (Zero Books,

2017) and Lawrence Wright, *Going Clear: Scientology, Hollywood, and the Prison of Belief* (Vintage, 2013).

6. Angela Nagle, "The Scourge of Self-Flagellating Politics," *Current Affairs* (February 22, 2017), https://www.currentaffairs.org/2017/01/the-scourge-of-self-flagellating-politics.

7. See Aaron James, *Assholes: A Theory* (Anchor, 2014).

8. I examine this at some length in Blacker, *Democratic Education Stretched Thin*.

9. Bradford Richardson, "Dave Rubin, lapsed progressive, explains why he left," *Washington Times* (February 6, 2017), https://www.washingtontimes.com/news/2017/feb/6/dave-rubin-lapsed-progressive-explains-why-he-left/.

10. See Tim Lott, "If leftwingers like me are condemned as rightwing, then what's left?" *The Guardian* (March 11, 2015), https://www.theguardian.com/commentisfree/2015/mar/11/mainstream-left-silencing-sympathetic-voices; and Nick Cohen, "Why I've finally given up on the left," *The Spectator* (September 19, 2015), https://www.spectator.co.uk/2015/09/why-ive-finally-given-up-on-the-left/.

11. "I'm very proud that I was a Goldwater Girl. And then my political beliefs changed over time." Scott Simon, "'Goldwater Girl': Putting Context To A Resurfaced Hillary Clinton Interview," *NPR Weekend Edition* (March 26, 2016), https://www.npr.org/2016/03/26/471958017/-goldwater-girl-putting-context-to-a-resurfaced-hillary-clinton-interview.

12. It is eye-opening to contrast the forgiveness regarding the later Wallace from an old school civil rights warrior and the raging sanctimony and call-out culture's implied perfectionism: "Congressman John Lewis, now the last surviving speaker of the 1963 March on Washington, wrote about Wallace in 1998, "our ability to forgive serves a higher moral purpose in our society. Through genuine repentance

and forgiveness, the soul of our nation is redeemed. George Wallace deserves to be remembered for his effort to redeem his soul and in so doing to mend the fabric of American society." Larry Provost, "The Redemption of George Wallace," *Townhall* (July 11, 2014), https://townhall.com/columnists/larryprovost/2014/07/11/the-redemption-of-george-wallace-n1860547.

13. Mark Fisher, *Capitalist Realism: Is There No Alternative?* (Zero Books, 2009).

14. A good example may be found in the reaction to pop star and leftist Russell Brand, who evidently is not clubbable enough for certain snobby Brits: David Sexton, "Brand is no revolutionary, just a reformed addict," *Evening Standard* (November 4, 2014), https://www.standard.co.uk/comment/comment/david-sexton-brand-is-no-revolutionary-just-a-reformed-addict-9838308.html. This snobbery exists on the left, too, as Mark Fisher memorably relates in "Exiting the Vampire Castle": "There's also a shocking but revealing aside where the individual casually refers to Brand's 'patchy education (and) the often wince-inducing vocab slips characteristic of the auto-didact,'" http://www.thenorthstar.info/2013/11/22/exiting-the-vampire-castle/.

15. kos, "Be happy for coal miners losing their health insurance. They're getting exactly what they voted for," *Daily Kos* (December 11, 2016), https://www.dailykos.com/stories/2016/12/12/1610198/-Be-happy-for-coal-miners-losing-their-health-insurance-They-re-getting-exactly-what-they-voted-for.

Chapter 4

1. *Selections from the Prison Notebooks, "Wave of Materialism" and "Crisis of Authority"* (International Publishers), (1971), 275-276.

2. Mill, *Utilitarianism* (1863), available at *Utilitarianism*

Resources, https://www.utilitarianism.com/mill1.htm.

3. A nineteenth-century Unitarian minister, Theodore Parker, is probably the originator of the phrase. See Quote Investigator, "The Arc of the Moral Universe Is Long, But It Bends Toward Justice," accessed April 2018, https://quoteinvestigator.com/2012/11/15/arc-of-universe/.

4. Mark Zuckerberg, "Building Global Community," *Facebook* (February 16, 2017), https://www.facebook.com/notes/mark-zuckerberg/building-global-community/10103508221158471/?pnref=story.

5. Friedrich Nietzsche, *Genealogy of Morals* [1887] (Vintage, 1989).

6. Friedrich Nietzsche, *Beyond Good and Evil: Prelude to a Philosophy of the Future* [1886] (Vintage, 1966), Preface.

7. Plato, *Phaedo*, 82e-83d, available at *The Internet Classics Archive*, http://classics.mit.edu/Plato/phaedo.html.

8. Kant, Immanuel, Prolegomena to Any Future Metaphysics [1772], (Hackett, 1977), Sec. 32.

9. "Symbols gather around the thing to be explained, understood, interpreted. The act of becoming conscious consists in the concentric grouping of symbols around the object, all circumscribing and describing the unknown from many sides." Erich Neumann, *The Origins and History of Consciousness* (Princeton University Press, 1954), 7.

10. Friedrich Nietzsche, *Will to Power* (Vintage 2011), Sec. 30a; sec. 125.

11. Ibid.

12. Ibid., sec. 215.

13. George Kateb, *The Inner Ocean: Individualism and Democratic Culture* (Cornell University Press, 1994), 128-130.

14. C. G. Jung, *Archetypes and the Collective Unconscious* in *The Collected Works of C. G. Jung: Volume 9, Part 1* (Princeton University Press, 1981).

15. Robert C. Solomon, "A More Severe Morality: Nietzsche's

Affirmative Ethics," in Yirmiahu Yovel, ed., *Nietzsche as Affirmative Thinker* (Martinus Nijhoff Publishers, 1983), 69.

16. *Thus Spoke Zarathustra,* trans. R.J. Hollingdale (Penguin Books, 1985).

17. Ibid., sec. 29.

18. Habermas, *Moral Consciousness and Communicative Action: Moral Consciousness and Communicative Action* (MIT Press, 2001).

19. Friedrich Nietzsche, "Letter to Overbeck," as quoted in R.J. Hollingdale, *Nietzsche: The Man and His Philosophy* (Cambridge University Press, 1999), 238.

20. Friedrich Nietzsche "Why I am so Clever," in *Ecco Homo* (Vintage, 1989), 236.

21. Friedrich Nietzsche, *The Case of Wagner* (Vintage, 1767), 163.

22. Nietzsche, *Will to Power.* His sister cut this aphorism out to make him palatable to the Nazis, see David Wrote, "'Criminal' manipulation of Nietzsche by sister to make him look anti-Semitic," *The Daily Telegraph* (January 19, 2010), https://www.telegraph.co.uk/news/worldnews/europe/germany/7018535/Criminal-manipulation-of-Nietzsche-by-sister-to-make-him-look-anti-Semitic.html.

23. Jefferson, *Notes on the State of Virginia.*

24. Peter Singer, *The Expanding Circle: Ethics, Evolution, and Moral Progress* (Princeton University Press, 2011).

25. Among his other works, see Frans de Waal, *Primates and Philosophers: How Morality Evolved* (Princeton University Press, 2006); Lee Glendinning, "Spanish parliament approves 'human rights' for apes," (June 26, 2008), https://www.theguardian.com/world/2008/jun/26/humanrights.animalwelfare

26. John Seed et al., *Thinking Like a Mountain: Toward a Council of All Beings* (New Society Publishers, 2007), 29. As quoted in Adrian Harris, "Ethical Approaches for Life on Earth: Deep Ecology, Ecofeminism or what ...?," *Exeter Philosophy Circle*

(January 11, 2017), available at https://www.thegreenfuse. org/papers/Exeter_Talk.pdf.

27. Lierre Keith, *The Vegetarian Myth: Food, Justice, and Sustainability* (PM Press, 2009).

28. "the best thing would be to just let nature seek its own balance, to let the people there just starve..." Quoted in George Bradford, "How Deep is Deep Ecology?," *The Anarchist Library*, accessed April 2018, https://theanarchistlibrary.org/ library/george-bradford-how-deep-is-deep-ecology#toc15.

29. Edmund Husserl, *Logical Investigations*, trans. J. N. Findlay (Routledge [1900/01; 2nd, revised edition 1913], 1973), sec. 1.

30. John and Roy Boulting, *Pastor Hall* (London 1939). Quoted in Tom Lawson, *The Church of England and the Holocaust: Christianity, Memory and Nazism*. (Boydell Press, 2006), 49.

31. *Groundwork of the Metaphysic of Morals*, 4:439. In Kant's moral philosophy this is essentially a thought experiment wherein everyone acted according to the formula of humanity and treated everyone else as ends in themselves. I am using the phrase more broadly to refer to any perfectionist utopian vision that can function as a regulative ideal and orient moral directionality.

32. Zygmunt Bauman, "Times of Interregnum," *Ethics and Global Politics* 5 (2012): 49-56.

33. Friedrich Nietzsche, *The Gay Science* (Vintage Books, 1974), sec. 125, 182.

34. *Manhattan*, directed by Woody Allen (Jack Rollins & Charles H. Joffe Productions, 1979). See also Christopher Middleton, *Middle Age: The Art of Living* (Routledge, 2009), 30. Middleton mentions Nietzsche also provides such a list in *Ecce Homo*: "his tea, cold water, dry air, a clear sky, the south of France, and northern Italy." Ibid.

35. Baruch Spinoza, *Ethics* (1677), available at Project Gutenberg, accessed April 2018, https://www.gutenberg. org/files/3800/3800-h/3800-h.htm.

36. Bron Taylor, *Dark Green Religion: Nature Spirituality and the Planetary Future* (University of California Press, 2010), 13.

37. Ibid., 16.

38. See for example Dave Jacke, *Edible Forest Gardens* (Chelsea Green Publishing, 2005).

39. Bart D. Ehrman, *The Triumph of Christianity: How a Forbidden Religion Swept the World* (Simon & Schuster, 2018).

40. Leviticus 18:22: "Thou shalt not lie with mankind, as with womankind: it is abomination." Leviticus 19:19: "neither shall a garment mingled of linen and woollen come upon thee."

41. Sean Illing, "Why Christian conservatives supported Trump—and why they might regret it," *Vox* (February 2, 2018), https://www.vox.com/2017/10/4/16346800/donald-trump-christian-right-conservative-clinton.

42. Michelle A. Vue, "Interview: David Platt on the American Dream, Radical Christianity," *The Christian Post* (May 15, 2010), http://www.christianpost.com/news/interviewdavid-platt-on-the-american-dream-radical-christianity-45161/.

43. David Platt, *Radical: Taking Back Your Faith from the American Dream* (Multnomah, 2010), 183. Platt's website is http://www.radical.net/.

44. Michel Houellebecq, *Submission: A Novel* (Picador, 2016).

45. Johann Gottfried von Herder, "Treatise on the Origin of Language," in *Philosophical Writings,* trans. Michael Forster (Cambridge University Press, 2002).

46. Andy Beta, "10 Things We Learned at Ta-Nehisi Coates' 'Black Panther' Cast Talk," *Rolling Stone* (February 28, 2018), https://www.rollingstone.com/movies/news/10-things-we-learned-ta-nehisi-coates-black-panther-w517210.

47. Carl R. Perkins, "The Alt-Right Has a New Hero and it's Black Panther," *International Policy Journal* (June 26, 2017), https://intpolicydigest.org/2017/06/26/the-alt-right-has-a-new-hero-and-it-s-black-panther/.

48. Adolph Reed Jr, "From Jenner to Dolezal: One Trans Good, the Other Not So Much."

49. A second wave of online outrage about Dolezal arose in 2018 as the result of the Netflix documentary "The Rachel Divide" (https://www.netflix.com/title/80149821); see also Diannah Watson, "New Rachel Dolezal Documentary Hits A Raw Nerve On Her Identity," *BlackAmericaEntertainmentWeb.com* (March 8, 2018), https://blackamericaweb.com/2018/03/08/new-rachel-dolezal-documentary-hits-a-raw-nerve-on-her-identity/.

50. "Race as biology is fiction, racism as a social problem is real: Anthropological and historical perspectives on the social construction of race." Audrey Smedley and Brian D. Smedley, *American Psychologist*, vol 60, no. 1 (January 2005): 16-26; Michael Yudell et al., "Taking race out of human genetics," *Science*, vol. 351, no. 6273 (February 5, 2016): 564-565.

51. Plotinus, *The Essential Plotinus* (Hackett, 1964), 74.

52. George Orwell, *1984* [1949], available at Project Gutenberg, http://gutenberg.net.au/ebooks01/0100021.txt.

53. Sabine Rewald, "Cubism," *Heilbrunn Timeline of Art History* (The Metropolitan Museum of Art, 2004), https://www.metmuseum.org/toah/hd/cube/hd_cube.htm.

54. Walter Gropius, *Manifesto and Programme of the Weimar State Bauhaus* (1919), available at https://www.bauhaus100.de/en/past/works/education/manifest-und-programm-des-staatlichen-bauhauses/.

55. Bauhaus was an early Nazi target due to its "internationalism." See Holland Carter, "First, They Came for the Art: 'Degenerate Art,' at Neue Galerie, Recalls Nazi Censorship," *New York Times* (March 13, 2014), https://www.nytimes.com/2014/03/14/arts/design/degenerate-art-at-neue-galerie-recalls-nazi-censorship.html. There were complexities though, as in the case of Bauhaus artist Herbert

Bayer: Alice Rawsthorn, "Exhibition Traces Bauhaus Luminary's Struggle With His Past" (January 7, 2014), https://www.nytimes.com/2014/01/08/arts/design/A-Bauhaus-Luminarys-Struggle-With-Past-as-Nazi-Propagandist.html.

56. McKenzie Wark, *The Beach Beneath the Street: The Everyday Life and Glorious Times of the Situationist International* (Verso, 2015).

57. Filippo Tomasso Marinetti, *The Futurist Manifesto* [1909], available at *Italian Futurism*, accessed April 2018, https://www.italianfuturism.org/manifestos/foundingmanifesto/.

58. Robert Hughes, *Nothing of Not Critical: Selected Essays on Art and Artists* (Penguin Books, 1987), 176.

59. Robert Hughes, *Rome: A Cultural, Visual, and Personal History* (Vintage, 2012), 400-404; and Philip McCouat, "The futurists declare war on pasta," *Journal of Art in Society* (2014), http://www.artinsociety.com/the-futurists-declare-war-on-pasta.html.

60. "Situationist Manifesto," (1960) trans. Fabian Thompsett, *Situationist International Online*, retrieved April 2018, http://www.cddc.vt.edu/sionline/si/manifesto.html.

61. Clive Philpott, "Manifesto I: Fluxus: Magazines, Manifestos, Multum in Parvo" (George Maciunas Foundation, Inc), retrieved April 2018, http://georgemaciunas.com/about/cv/manifesto-i/.

62. *Do Androids Dream of Electric Sheep?* (Del Rey, 1996).

63. Lynn White, *Medieval Technology and Social Change* (Oxford University Press, 1966).

64. John Horgan, "The Singularity and the Neural Code," *Scientific American* (March 22, 2016), https://blogs.scientificamerican.com/cross-check/the-singularity-and-the-neural-code/.

65. James Lovelock, *Gaia: A New Look at Life on Earth* (Oxford University Press, 1979), 58.

66. James Lovelock, *Rough Ride to the Future* (The Overlook

Press, 2014), 161.

67. Ray Kurzweil, *The Singularity is Near: When Humans Transcend Biology* (Penguin Books, 2006).

68. John Horgan, "The Consciousness Conundrum," *IEEE Spectrum* (June 1, 2008), https://spectrum.ieee.org/ biomedical/imaging/the-consciousness-conundrum.

69. Emerging Technology from the arXiv, "First Evidence That Online Dating Is Changing the Nature of Society," *MIT Technology Review* (October 10, 2017), https://www.technologyreview.com/s/609091/first-evidence-that-online-dating-is-changing-the-nature-of-society/.

70. Vanessa McMains, "Hallucinogenic drug found in 'magic mushrooms' eases depression, anxiety in people with life-threatening cancer," *Johns Hopkins Medicine* (December 1, 2016), https://hub.jhu.edu/2016/12/01/hallucinogen-treats -cancer-depression-anxiety/; Jose Carlos Buoso et al., "Serotonergic psychedelics and personality: A systematic review of contemporary research," *Neuroscience & Biobehavioral Reviews*, vol. 87 (April 2008): 118-132. See also Rick Strassman, *DMT: The Spirit Molecule* (Park Street Press, 2001).

71. Reynaldo Anderson, "Afrofuturism 2.0 & the Black Speculative Art Movement: Notes on a Manifesto," *Obsidian*, 240-271, retrieved April 2018, https://monoskop.org/images/a/a9/Anderson_ Reynaldo_2016_Afrofuturism_2.0_and_the_Black_ Speculative_Arts_Movement.pdf.

72. Ibid.

73. Ibid.

74. Candice Frederick, "Art, Futurism, and the Black Imagination," *New York Public Library* (September 29, 2015), https://www.nypl.org/blog/2015/09/29/art-futurism-black-imagination. The exhibit, *Unveiling Visions: The*

Alchemy of the Black Imagination (New York Public Library, Shomburg Center for Research in Black Culture, October 1, 2015-January 16, 2016), may be accessed here: https://www.nypl.org/events/exhibitions/unveiling-visions

75. Kashmira Gander, "Ayahuasca: The Lawyer Fighting for those who take the Hallucinogenic Drug for Religious Reasons," *The Independent* (February 28, 2017), https://www.independent.co.uk/life-style/ayahuasca-lawyer-j-hamilton-hudson-hallucinogenic-drug-religious-reasons-south-america-amazon-tribes-a7603341.html. The same could be said of the more established (in the US) ritual peyote usage in the Native American Church. See John Horgan, "Tripping on Peyote in Navajo Nation," *Scientific American* (July 5, 2017), https://blogs.scientificamerican.com/cross-check/tripping-on-peyote-in-navajo-nation/. See also *Employment Division v. Smith*, 497 US 872 (1990) (Oregon's interest in regulating controlled substances trumps peyote users' free exercise) but also *Gonzales v. O Centro Espirita Beneficente Uniao do Vegetal*, 546 US 418 (2006) (under the Religious Freedom Restoration Act the government fails to establish a compelling interest in prohibiting religious ceremonial use of ayahuasca).

76. Roland R. Griffiths, "Psilocybin produces substantial and sustained decreases in depression and anxiety in patients with life-threatening cancer: A randomized double-blind trial," *Journal of Psychopharmacology*, vol. 30, no. 12 (November 30, 2016): 1181-1197; see also Vanessa McMains, "Study explores the enduring positive, negative consequences of ingesting 'magic mushrooms,'" *Johns Hopkins Medicine* (January 4, 2017), https://hub.jhu.edu/2017/01/04/bad-trips-mushrooms/.

77. See Peter Wohlleben, *The Hidden Life of Trees: What They Feel, How They Communicate—Discoveries from a Secret World* (Greystone Books, 2016).

78. A helpful and sober example of this is Sigal R. Ben-Porath, *Free Speech on Campus* (University of Pennsylvania Press, 2017).

79. Sam Gindin, "Beyond Social Movement Unionism," *Jacobin* (August 17, 2016), https://www.jacobinmag.com/2016/08/beyond-social-movement-unionism.

80. Steven Greenhouse, "The West Virginia Teacher Strike Was Just the Start," *New York Times* (March 7, 2018), https://www.nytimes.com/2018/03/07/opinion/teachers-west-virginia-strike.html; Pedro Noguera, "What's at Stake in the Teachers' Strikes?" *The Nation* (April 19, 2018), https://www.thenation.com/article/whats-at-stake-for-striking-teachers/.

81. Jedediah Purdy, "A New Struggle Coming: On the teachers' strike in West Virginia," *Nplusonemag.com* (March 5, 2018), https://nplusonemag.com/online-only/online-only/a-new-struggle-coming/.

82. Chris McGreal, "'The S-word': how young Americans fell in love with socialism," *The Guardian* (September 2, 2017), https://www.theguardian.com/us-news/2017/sep/02/socialism-young-americans-bernie-sanders.

Chapter 5

1. Isabelle Stengers, *In Catastrophic Times: Resisting the Coming Barbarism* (Open Humanities Press, 2015), available at http://modesofexistence.org/isabelle-stengers-the-intrusion-of-gaia/.

2. James Lovelock, *Gaia*; and Lynn Margulis, *Symbiotic Planet: A New Look at Evolution* (Basic Books, 1998).

3. Lovelock, *Gaia*, xv.

4. Michael Gross "How Life Shaped Earth," *Current Biology* 25 (October 5, 2015): 845-875, http://www.cell.com/current-biology/pdf/S0960-9822(15)01090-8.pdf; Robert M. Hazen, *The Story of Earth: The First 4.5 Billion Years, from Stardust to Living Planet* (Penguin, 2013).

5. Norman Sleep and Dennis Bird, "Niches of the pre-photosynthetic biosphere and geologic preservation of Earth's earliest ecology," *Geobiology* (2007): 101-117; Norman Sleep, "Biogeochemistry: Oxygenating the atmosphere," *Nature* 410 (March 15, 2001): 317–319.

6. Lovelock, *Gaia*, 99.

7. Lovelock, *Rough Ride to the Future*, 81.

8. Lovelock, *Gaia*, 10.

9. David Grinspoon, *Earth in Human Hands: Shaping Our Planet's Future* (Grand Central Publishing, 2016).

10. Lovelock, *Gaia*, 9-10.

11. Toby Tyrrell, *On Gaia: A Critical Investigation of the Relationship between Life and Earth* (Princeton University Press, 2013).

12. Lynn Margulis, *Symbiotic Planet: A New Look at Evolution* (Basic Books, 1998), 123.

13. Ibid.

14. Lovelock, *Gaia*, 31.

15. Bruno Latour, *Politics of Nature: How to Bring the Sciences into Democracy* (Harvard University Press, 2004), 155–156.

16. Timothy Morton, *Hyperobjects: Philosophy and Ecology after the End of the World* (University of Minnesota Press, 2013).

17. Eleanor Ainge Roy, "New Zealand gives Mount Taranaki same legal rights as a person," *The Guardian* (December 22, 2017), https://www.theguardian.com/world/2017/dec/22/new-zealand-gives-mount-taranaki-same-legal-rights-as-a-person.

18. Mary Midgley, *Gaia: The Next Big Idea* (Demos, 2001), 14. The Plato quote is from *Timaeus* 31a, available at *The Internet Classics Archive*, http://classics.mit.edu/Plato/timaeus.html.

19. Roy, "New Zealand gives Mount Taranaki same legal rights as a person."

20. Margulis, *Symbiotic Planet*, 128.

21. Bruno Latour, "From Realpolitik to *Dingpolitik* or How to Make Things Public in Making Things Public." In Bruno

Latour and Peter Weibel, eds., *Making Things Public: Atmospheres of Democracy* (Karlsruhe: ZKM Center for Art and Media, 2005), 14, as quoted in Massimiliano Simon, "The Parliament of Things and the Anthropocene: How to Listen to 'Quasi-Objects'" *Techné: Research in Philosophy and Technology,* vol. 21, nos. 2–3 (2017), 2.

22. Stengers, *In Catastrophic Times.*
23. Grinspoon, *Earth in Human Hands,* 78.
24. Descartes, *Meditations* (Hackett, 2006). For a further corrective appreciation of Descartes, see Blacker, *Democratic Education Stretched Thin,* 128-140.
25. Richard Rorty, *Consequences of Pragmatism* (University of Minnesota Press, 1982), xxxix.
26. *On the Nature of Things,* trans. Anthony M. Esolen (Cengage Learning, 1995), 166.
27. Robinson Jeffers, "Carmel Point," in *The Selected Poetry of Robinson Jeffers* (Stanford University Press, 2002).
28. Robinson Jeffers, "Vulture," in *The Selected Poetry of Robinson Jeffers.*
29. Paulette Jiles, *News of the World* (William Morrow, 2017), 64.
30. (University of California Press, 2017), 1. The quote from Murphy-Ellis is from Usnea, "My Name is Emma Murphy-Ellis. . .And I Support Sabotage," *Earth First! Journal* (2011), available at https://drstevebest.wordpress.com/2012/08/10/my-name-is-emma-murphy-ellis-and-i-support-sabotage/.
31. Pike, *For the Wild,* 1.
32. See Andrew Linzey, *Creatures of the Same God: Explorations in Animal Theology* (Winchester University Press, 2007), as quoted in Alisa Aaltola, *Animal Suffering: Philosophy and Culture* (New York: Palgrave Macmillan, 2012), 132.
33. Pike, *For the Wild,* 87-88.
34. Ibid., 21-22.
35. Ibid., 212.
36. Ibid., 115.

37. Jeff Luers, "How I Became an Eco-Warrior," as quoted in Pike, *For the Wild,* 106.

38. Pike, *For the Wild,* 233-235.

39. Gregory Dicum, "An interview with jailed 'eco-terrorist' Jeffrey Luers," *Grist* (May 4, 2006), https://grist.org/article/dicum1/.

40. North American Earth Liberation Prisoners Support Network, retrieved April 2018, http://www.ecoprisoners.org/.

41. Pike, *For the Wild,* 223-224.

42. Bill Mollison, *Introduction to permaculture* (Tagari, 1991).

43. Mark Hoffman, "Permaculture Examples in Our Garden," retrieved April 2018, http://greenhousebed.com/Permaculture/permaculture_examples.htm.

44. Debal Deb, "Sacred Groves of West Bengal: A Model of Community Forest Management," *SSRN,* Elsevier (March 3, 2014), https://papers.ssrn.com/sol3/papers.cfm?abstract_id=2403540.

45. Deb, *Beyond Developmentality: Constructing Inclusive Freedom and Sustainability* (London: Earthscan, 2009), 345.

46. Ibid.

47. Ibid., 265.

48. Estimates of neopagans and Wiccans et al. in the US seem to be all over the place, depending upon definitions. Ditto for indigenous peoples actually practicing their traditional indigenous religions. Nobody seems to claim there are over a few million, though. See Paul Heelas, *The New Age Movement: The Celebration of the Self and the Sacralization of Modernity* (Wiley-Blackwell, 1996), 218.

49. See Catherine L. Albanese, *Nature Religion in America: From the Algonkian Indians to the New Age* (University of Chicago Press).

Epilogue

1. George Carlin, *Brain Droppings* (Hachette, 1998), 100.
2. adrienne marie brown, *Emergent Strategy*, 42.
3. Martin Heidegger, *Being and Time* (Harper, 2008), 32.
4. Mark Fisher, "Exiting the Vampire Castle."
5. George Monbiot, *Out of the Wreckage: A New Politics for an Age of Crisis* (Verso, 2017), 1.
6. Timothy Morton, *Humankind: Solidarity with Nonhuman People* (Verso, 2017), 3.

Acknowledgments

I am very grateful to the individuals and audiences who have supported this work. Notable among these are Kevin McDonough and the students and faculty at the Department of Integrated Studies at McGill University, and Chris Higgins and the students, faculty and audiences at the Department of Education, Organization and Leadership, the Center for Advanced Study, and the Unit for Criticism and Interpretive Theory at the University of Illinois at Urbana-Champaign. This work was also supported indirectly through the research portion of my employment duties at the University of Delaware and I am grateful to my home institution's continuing commitment to such scholarship—something that is not to be taken for granted. I received tremendous help in the form of manuscript commentary from numerous friends and colleagues, including Aideen Murphy, Roxanne Desforges, Chris Phillips, Mike Watson, Quentin Wheeler-Bell, Terri Wilson, Carolyn Cohen, Dana Simone, Emile Bosjean, Matthew Charles, Marcia Blacker, Marion Monguillon, Marianna Papastefanou, and Alpesh Maisuria. Any remaining errors are of course mine alone despite these individuals' valiant efforts to save me from them. At Zero Books I am very happy to thank John Hunt, Ashley Frawley, Nick Welch, Stuart Davies, Trevor Greenfield, Doug Lain and John Romans for their help at various stages. It was a pleasure to work with them.

Index

Zero Books

CULTURE, SOCIETY & POLITICS

Contemporary culture has eliminated the concept and public figure of the intellectual. A cretinous anti-intellectualism presides, cheer-led by hacks in the pay of multinational corporations who reassure their bored readers that there is no need to rouse themselves from their stupor. Zer0 Books knows that another kind of discourse – intellectual without being academic, popular without being populist – is not only possible: it is already flourishing. Zer0 is convinced that in the unthinking, blandly consensual culture in which we live, critical and engaged theoretical reflection is more important than ever before. If you have enjoyed this book, why not tell other readers by posting a review on your preferred book site.

Recent bestsellers from Zero Books are:

In the Dust of This Planet
Horror of Philosophy vol. 1
Eugene Thacker
In the first of a series of three books on the Horror of Philosophy,
In the Dust of This Planet offers the genre of horror as a way of
thinking about the unthinkable.
Paperback: 978-1-84694-676-9 ebook: 978-1-78099-010-1

Capitalist Realism
Is there no alternative?
Mark Fisher
An analysis of the ways in which capitalism has presented itself
as the only realistic political-economic system.
Paperback: 978-1-84694-317-1 ebook: 978-1-78099-734-6

Rebel Rebel
Chris O'Leary
David Bowie: every single song. Everything you want to know,
everything you didn't know.
Paperback: 978-1-78099-244-0 ebook: 978-1-78099-713-1

Cartographies of the Absolute
Alberto Toscano, Jeff Kinkle
An aesthetics of the economy for the twenty-first century.
Paperback: 978-1-78099-275-4 ebook: 978-1-78279-973-3

Malign Velocities
Accelerationism and Capitalism
Benjamin Noys
Long listed for the Bread and Roses Prize 2015, *Malign Velocities* argues against the need for speed, tracking acceleration as the symptom of the ongoing crises of capitalism.
Paperback: 978-1-78279-300-7 ebook: 978-1-78279-299-4

Meat Market
Female Flesh under Capitalism
Laurie Penny
A feminist dissection of women's bodies as the fleshy fulcrum of capitalist cannibalism, whereby women are both consumers and consumed.
Paperback: 978-1-84694-521-2 ebook: 978-1-84694-782-7

Poor but Sexy
Culture Clashes in Europe East and West
Agata Pyzik
How the East stayed East and the West stayed West.
Paperback: 978-1-78099-394-2 ebook: 978-1-78099-395-9

Romeo and Juliet in Palestine
Teaching Under Occupation
Tom Sperlinger
Life in the West Bank, the nature of pedagogy and the role of a university under occupation.
Paperback: 978-1-78279-637-4 ebook: 978-1-78279-636-7

Sweetening the Pill
or How We Got Hooked on Hormonal Birth Control
Holly Grigg-Spall
Has contraception liberated or oppressed women? *Sweetening the Pill* breaks the silence on the dark side of hormonal contraception.
Paperback: 978-1-78099-607-3 ebook: 978-1-78099-608-0

Why Are We The Good Guys?
Reclaiming your Mind from the Delusions of Propaganda
David Cromwell
A provocative challenge to the standard ideology that Western power is a benevolent force in the world.
Paperback: 978-1-78099-365-2 ebook: 978-1-78099-366-9

Readers of ebooks can buy or view any of these bestsellers by clicking on the live link in the title. Most titles are published in paperback and as an ebook. Paperbacks are available in traditional bookshops. Both print and ebook formats are available online.

Find more titles and sign up to our readers' newsletter at http://www.johnhuntpublishing.com/culture-and-politics

Follow us on Facebook
at https://www.facebook.com/ZeroBooks

and Twitter at https://twitter.com/Zer0Books